Praise for *Tomatoes*

Jean-Martin Fortier's tomato guide is full of tips that will save new growers years of wasted effort. Mastering tomatoes—with this book—is a great place to start your farming career.
— Ben Hartman, author, *The Lean Farm* and *The Lean Micro Farm*

Tomatoes are a profitable crop for market growers and worthy of mastery. Jean-Martin shares his experience and wisdom on this important crop in an accessible way for new and advanced growers.
— Zach Loeks, edible cco-system designer, and author, *The Garden Tool Handbook*

The contents of this book belie its small size; the book is packed with solid information, which, along with the layout and illustrations, make for an engaging and very pleasant read. This Grower's Guide has something for everyone, from seasoned gardeners and small farmers to newbies. Read it. Enjoy it. Learn from it.
— Lee Reich, PhD, scientist, farmdener (farmer/gardener hybrid), and author, *Growing Figs in Cold Climates*

Everything you need to know to grow amazing tomatoes.
— David R. Montgomery, co-author, *What Your Food Ate* and *The Hidden Half of Nature*

I was thrilled with this book. Tomatoes are so varied and there's much folklore about them. I love the illustrations and pored over them. The techniques and strategies are useful and appropriate. The information is well organised and the language is clear. It is such a useful addition to home gardeners and commercial growers.
—Rowe Morrow, Blue Mountains Permaculture Institute, author, *Earth Restorer's Guide to Permaculture*

Market gardening innovator Jean-Martin Fortier draws on his decades of experience in a difficult climate to share his strategies and techniques for producing high yields of sweet, juicy tomatoes.
—Darrell E. Frey, author, *Bioshelter Market Garden*, and co-author, *The Food Forest Handbook*

It's always fun and informative to read a seasoned grower's tips and tricks for any crop, and tomatoes are an important crop for market growers, with lots of steps getting from seed to market. I know Jean-Martin has spent a lot of time working with tomato production, so it's great to see his met
—Josh Volk, Slow Hand Farm, author, *Comp*

Praise for the Grower's Guides
from the Market Gardener series

Jean-Martin Fortier has done it again! Incredibly beautiful, this series of pragmatic and professional instructional manuals makes it easy for anyone to start market gardening at a professional and regenerative level with a tailored focus on each aspect.
—Matt Powers, author, *Regenerative Soil* and *Regenerative Soil Microscopy*

Applying a market gardening mindset to your home garden will improve your yields and greatly reduce the amount of work involved. To do this, look no further than this series from Jean-Martin Fortier who is a pioneer in regenerative, biointensive, and organic market gardening.
—Rob and Michelle Avis, Verge Permaculture / 5th World, co-authors, *Building Your Permaculture Property*

GROWER'S GUIDES
Jean-Martin Fortier
FROM THE MARKET GARDENER

Tomatoes
A Grower's Guide

TRANSLATED BY LAURIE BENNETT
EDITED BY PIERRE NESSMANN
ILLUSTRATIONS BY FLORE AVRAM

Copyright © 2025 by Jean-Martin Fortier. All rights reserved.
Translated by Laurie Bennett. Cover design by Diane McIntosh.
Printed in Canada. First printing March, 2025.

© Delachaux et Niestlé, Paris, 2023 First published in France
under the title: *Les tomates. Les guides du jardinier-maraîcher*
Jean-Martin Fortier, Flore Avram.

The author and publisher disclaim all responsibility for any liability,
loss, or risk that may be associated with the application of any of the
contents of this book.

Inquiries regarding requests to reprint all or part of *Tomatoes* should
be addressed to New Society Publishers at the address below. To
order directly from the publishers, please call 250-247-9737 or order
online at www.newsociety.com.

Any other inquiries can be directed by mail to:
New Society Publishers
P.O. Box 189, Gabriola Island, BC
V0R 1X0, Canada
(250) 247-9737

LIBRARY AND ARCHIVES CANADA CATALOGUING IN PUBLICATION

Title: Tomatoes / Jean-Martin Fortier ; translated by Laurie Bennett ;
    edited by Pierre Nessmann ; illustrations by Flore Avram.
Other titles: Tomates. English
Names: Fortier, Jean-Martin, author
Description: Series statement: Grower's guides from the market
    gardener ; 1 | Translation of: Les tomates. | Includes bibliographical
    references. | In English, translated from the French.
Identifiers: Canadiana (print) 20240510623 |
    Canadiana (ebook) 2024051064X | ISBN 9781774060056
(softcover) | ISBN 9781550927986 (PDF) |
    ISBN 9781771423946 (EPUB)
Subjects: LCSH: Tomatoes.
Classification: LCC SB349 .F6713 2025 | DDC 635/.642—dc23

New Society Publishers' mission is to publish books that contribute
in fundamental ways to building an ecologically sustainable and
just society, and to do so with the least possible impact on the
environment, in a manner that models this vision.

Creating a future where humans live in harmony with nature and with each other

Founded by Jean-Martin Fortier, the Market Gardener Institute is committed to inspiring and supporting new organic growers at every stage of their journey. Our mission is to equip them with the essential technical skills needed to thrive in their vital agricultural work.

Our vision is to multiply the number of organic, regenerative farms around the world and create a future where humans live in harmony with nature and each other.

www.themarketgardener.com

Presenting the collection

Grower's Guides from the Market Gardener

Hi!

I am delighted to bring you this new collection of practical guides. The advice you'll find in these books is based on working methods I developed on my own microfarm and refined over the last two decades. While plenty of these concepts are not new and were passed on to me by different mentors through the years, many other ideas stem from my own farming experience. I am sure you'll come across a number of tips and tricks that are innovative, proven, and easy to implement.

Whether you are a home gardener, hobby farmer, new market gardener, or an experienced farmer looking to transition to more intensive growing on smaller plots, you will find everything you need to take your horticultural practices even further.

Wishing you success and happiness in your agricultural adventures!

Jean-Martin Fortier, market gardener in Saint-Armand, Quebec

Contents

Introduction: A Few Words About
My Background 1
What Is the Market Gardener Method? ... 4
Using This Guide to Successfully
Grow Tomatoes 8
Tomatoes: The Essentials 9

**JEAN-MARTIN FORTIER'S
20 FAVORITE TOMATOES** 15

GROWING SEEDLINGS 37

Seeding 38
Seeding into Pots 39
Seeding into Open Flats 42
Seeding in Plug Flats 45

Potting Up 50
Potting Up: From Open Flats to Pots ... 51
Potting Up: From Plug Flats to Pots ... 54

PLANTING AND EARLY CARE 57

Preparing the Soil 58
Permanent Beds 59
Planter Boxes and Garden Mounds 62

Planting 64
Planting in Greenhouses or Tunnels 65
Planting in the Field 67

TRAINING 70
Training Field and Garden Tomatoes 71
Trelllising Under Shelter 74

Mulching 76
Organic Mulch 77
Synthetic Mulch 78

**MAINTENANCE, HARVEST,
AND STORAGE** 81

Irrigation 82
Irrigation with a Watering
Can and Hose 83
Drip Irrigation 86

Pruning 88
Pruning Suckers 89
Pruning for Two Leaders 91
Pruning Lower Leaves 93

Fertilization 94
Mid-Season Fertilization
or Side Dressing 95

Harvest 100
Harvesting Tomatoes 101
Temporarily Storing Tomatoes 103
Storing Tomatoes at Home 104

Tomato Enemies 107
Diseases 108
Insect Pests and Parasites 112
Physiological Disorders 116

Acknowledgments 120
About New Society Publshers 120

Introduction: A Few Words About My Background

Drawing on principles from agroecology, permaculture, and entrepreneurship, I champion a modern form of nonmechanized farming, carried out on a human scale.

On a human scale means feeding many local families, while respecting the human and natural ecosystems in which we operate.

On a human scale means allowing market gardeners to make a decent living from their work, to run their businesses as they see fit, and to give themselves more time off than conventional farmers.

On a human scale means evolving through the use of technology but especially by relying on people and their skills and knowledge.

From Organic Farms...

I studied agroecology at McGill University's School of Environment in Montréal, where I met my wife and business partner, Maude-Hélène Desroches. At the time, we were both looking to create a new model for farming, one that would have a positive environmental impact. After graduation, we spent two years in New Mexico, USA, working on an organic farm and learning to be market gardeners.

Our microfarming aspirations were later fueled by a trip to Cuba where we spent time on *organopónicos*, fascinating urban farms that were established during the American embargo. During that era, after the fall of the USSR, the country developed a biointensive and urban agricultural model to ensure food security for the island's residents.

...to a Family-Run Microfarm

Back in Quebec in 2004, we acquired a small plot of 10 acres in Saint-Armand, in the scenic Eastern Townships. On this land, we experimented with our innovative approach to market gardening, which especially drew from the work of Eliot Coleman, an American market gardener who has been highly influential in the world of organic microfarming.

We built a 2-acre market garden, Les Jardins de la Grelinette, where we were able to test the first iterations of my method, now called the Market Gardener Method. It consists of crop rotation, the near-exclusive use of hand tools, organic growing practices, and shorter marketing channels, with direct sales made through CSA boxes and farmers' markets. At Les Jardins de la Grelinette, Maude-Hélène and I both worked full-time, and hired two farm workers (one full-time and the other part-time) to help with harvests.

Making 2 Acres Profitable

Success came quickly, both in terms of harvests and direct sales. After bringing in $33,000 in our first year, we earned twice that in the following year, and more than $110,000 in our third year of operation.

We were thus able to earn a living as market gardeners from almost the very beginning. Since then, our farm has continued to feed more than 200 families every year, offering roughly 40 types of vegetables, all grown on just 2 acres. Over the years, our harvests expanded and sales continued to increase. Eight years after starting the farm, I presented this farming model in a practical guide called *The Market Gardener* in 2014. The book was an instant success—over 250,000 copies have now been sold, and it has been translated into nine languages.

In 2015, with the support of a generous patron, I founded Ferme des Quatre-Temps in Hemmingford, Québec, with the vision of creating a model for the future of ecological agriculture. On this 160-acre farm, we established a polyculture system in a closed-loop cycle, raising pasture-fed cattle, pigs, and hens, alongside a culinary laboratory. At the heart of the farm, 7.5 acres were dedicated to a market garden, where we applied the growing methods developed at Les Jardins de la Grelinette. It is here that I teach my apprentices the principles of productive and profitable market gardening.

The project was featured in a TV show called *Les fermiers*, which follows the evolution of Ferme des Quatre-Temps and its apprentices, who later start their own farms in front of the cameras. The show was a hit in Quebec and is now available on TV5 Monde and Apple TV.

In parallel, I worked to expand my methods to reach a broader, global audience. In 2018, we launched the Market Gardener Masterclass, a fully online course now available in over 90 countries. To further support this initiative, I founded the Market Gardener Institute with a clear mission: to educate the next generation of growers by equipping them with the knowledge, skills, and resources needed to become leaders in the organic farming movement.

The Institute has two key objectives: to teach best practices in market gardening techniques and growing methods, and to demonstrate that small-scale farming worldwide can not only be ecological but also productive and profitable. On a global scale, it's the number of farms, not their size, that holds the key to feeding the world.

Inspiring Change

My ambition is to drive meaningful change in society by promoting a way of farming that honors nature, supports communities, and empowers local farmers. I believe in a decentralized farming model, built farm by farm, as the foundation for a truly sustainable and resilient food system.

Since 2020, I have proudly served as an ambassador for the prestigious Rodale Institute, which researches regenerative organic farming practices in the United States and beyond. I am also honored to be the ambassador for Growers and Co., a company that develops tools and apparel for new organic growers.

What Is the Market Gardener Method?

While my approach may seem innovative, it is founded on practices that were first developed by nineteenth-century Parisian gardeners, who fed more than two million people through a network of thousands of market gardens—precursors to our modern-day microfarms—within the city of Paris.

These market gardeners applied remarkable ingenuity, skills, and knowledge to meet the increasing food demands of a city in the midst of urbanization and demographic expansion. They achieved this through organic, nonmechanized agriculture. From the mid-eighteenth century to the twentieth century, many books were written about the innovative practices of these market gardeners, whose technical feats were admired throughout Europe. But with the advent of modern practices, much of this know-how was relegated to the past.

As a result of mechanization, the advent of agronomic science, and improved refrigeration and transport that brought in fresh and inexpensive food grown abroad, farms grew in size, became less diversified, and took on a more technological focus—a trend that continues today.

Fortunately, these inspiring models led to the development of horticultural methods that have endured, and with the same objective: to grow sustainably, by maximizing vegetable yields without degrading soil quality. We now use the term "biointensive" to describe these methods. Unlike extensive agricultural operations, they continue to work on a human scale and offer farmers the opportunity to use little mechanization. Despite what some may believe, this approach is also profitable.

By working on only small plots of land, market gardeners can keep start-up investments to a minimum, compared to the funds needed for a conventional farm. Biointensive farmers also require a smaller workforce, doing the work themselves with the help of just a few employees. They also sell their produce directly to customers, avoiding commissions to intermediaries. These three factors allow market gardeners to start generating profits quickly.

Still, it's important to remember that working the land is never easy. While market gardeners can make a good living with this method, the first seasons are time-consuming and require a significant workload and financial investment. In this profession, nothing comes easy, and every dollar you earn is the fruit of your labor, the result of your organizational skills. That's why I always tell my apprentices to learn how to work smarter, not harder.

From a financial perspective, market gardeners should plan to start with an investment of $50,000 to $150,000, depending on whether certain assets are already available—such as a building that can be converted, access to abundant water, electricity, natural gas, or a vehicle. This amount does not include the cost of purchasing land, which can be amortized over 20 years, if needed. Renting is also an option that can prove very profitable, especially when the farm is located near a city or an affluent municipality, where land is expensive.

Regardless of experience and preparation, the first years of market gardening will be intense. Opening new ground, constructing greenhouses and tunnels, and setting up infrastructure (irrigation, washing and packing stations, nurseries, etc.) all take extra time and effort. However, once this phase is complete, market gardeners who have mastered their craft can do more than just make a living off a few acres—they can earn a very decent living.

This leads to another key principle I teach: your farm should work for you, not the other way around. Profitable and productive farming is possible, but you need to set it up for success.

Using This Guide to Successfully Grow Tomatoes

Sun-drenched, healthy, flavorful tomatoes are, without a doubt, a signature crop for any good market gardener. The truth is, tomatoes are easy to grow, and as proof, many home gardeners grow bountiful and exquisitely tasty tomato crops. These harvests become precious treasures that growers proudly show off and share over meals with friends or family.

Of course, not every harvest is a success, and this is where a little expertise can make all the difference. To grow healthy tomatoes, you must be familiar with the different varieties, know their flavor profiles, and understand their particular horticultural needs. It is also important to manage certain processes to ensure a good harvest: seeding tomatoes and caring for seedlings, organic fertilization and proper irrigation, as well as smart disease and pest management. In short, you need to develop a good grasp of myriad details.

Each of these concepts is described in depth in this practical guide. You'll find the essentials needed to successfully grow your own tomatoes, as well as tips and tricks so that you can harvest exceptional fruits.

While this book was written for all home gardeners looking to grow better tomato crops, some sections are intended for my fellow market gardeners, whether they are experienced or in the process of starting their own microfarm!

Wishing you a resounding success in your gardening adventures.

Tomatoes: The Essentials

Tomatoes belong to the Solanaceae family, which also includes potato and eggplant. This type of plant is referred to as a "tender perennial," and, around the world, there are roughly 2,700 known Solanaceae species. While they are perennial plants in their native tropical climate, they are grown as annuals in more temperate regions. Tomato fruits come in a wide array of colors, shapes, and sizes, from plants that range from under 1 foot (30 cm) to over 6 feet (2 m) high. Both the plant and the fruit are called "tomato."

A Brief History of Tomatoes

All varieties of garden tomatoes belong to the same species, *Solanum lycopersicum*, which is native to western South America. The plant's endemic range extends from the equator to northern Chile. The Aztecs were the first to cultivate and domesticate several varieties, and they are credited with creating cherry tomatoes, or *Solanum lycopersicum cerasiforme*. It is thought that large tomato varieties were likely developed earlier through a selection process dating back to at least 500 BC. The Incas gave the plant its name: *tomalt*.

In the 16th century, when Spanish conquistadors in Mexico discovered this plant, as well as potatoes and squash, they brought it back to the Iberian Peninsula. The tomato then made its way to Italy, which at that time belonged to the Spanish crown where it earned the named *pomodoro*, meaning "golden apple," because the fruit was often yellow. Initially, it was used only as an ornamental plant and shunned by scientists who considered it nonedible. Worse, still, because it belonged to the same family as mandrakes, it was said to have aphrodisiac virtues and was given the name "love apple," rendering it suspect in the eyes of the Church. Though the word *tomate*, French for "tomato," had been used since 1532, it did not enter the Académie française dictionary until 1832.

Even though peasants had been growing and eating tomatoes since the plant first arrived on Spanish soil, it was not classified as an edible fruit until a 1750 publication by Carl Linnaeus. In 1778, the French seed company Vilmorin-Andrieux, which had long considered tomatoes to be a decorative plant, introduced a few varieties into their vegetable catalog. Towards the end of the 18th century, the tomato was brought to northern Europe. Parisian market gardeners started growing it to meet demand, and tomato cultivation spread to Russia, where it was produced in large quantities. Eventually, the tomato returned to the American continent in the early 19th century, thanks to Thomas Jefferson, who had discovered it during a trip to France. The last remaining fears surrounding its edibility abated when Dr. John Bennett published a glowing article about tomatoes in the *New York Times*.

Seed companies played an important role in encouraging home gardeners to grow tomatoes. In the 20th century, they created new varieties to meet standards set by the food industry, which preferred highly productive plants and uniform fruits. In 1971, at the behest of the Heinz company, seed producers developed the first cultivar intended to be machine harvested.

Today, tomatoes are grown in 170 countries, on 11.6 million acres of agricultural land, with 177 million tons of fruit harvested every year. The top three tomato producers and consumers are, in order, China, India, and the United States.

There is a clear difference between tomatoes grown on industrial farms and tomatoes grown in gardens. The industrial sector grows cultivars that favor high yields over flavor, taste, and aroma. These plants have better disease resistance and produce fruit year-round that keeps longer thanks to the RIN (ripening inhibitor) gene, which is responsible for their bland flavor and mealy texture. In comparison, home gardeners grow tomatoes with a rich genetic heritage, a delight for all tomato lovers, whether they are growing or just eating them!

Characteristics of Tomato Varieties

In its natural environment, the tomato plant is a vine that grows indefinitely. When cultivated in temperate climates (with short growing periods) and under market gardening needs (for efficient harvests), the tomato plant is a bush made up of one main stem and secondary branches, called "suckers" because they divert resources at the expense of the main stem and its flower clusters; this is known as sympodial growth. Once three leaf stems have developed, buds appear in the leaf axil, where the leaf stem meets the main stem, and then grow into suckers. The plants also have an alternate leaf arrangement, which means that leaves grow on alternating sides of the stem. Because the flowers are hermaphroditic, carrying both female and male organs, they can self-pollinate. In general, seeds harvested from tomatoes can be used to grow the same variety again. Hybrid varieties, however, are typically identified by the designation "F1", and their offspring will not be identical; with seeds collected from a hybrid, you never know what kind of tomato plant you'll get!

Tomato roots are dense and thick in the first few inches and can reach a depth of 3.5 feet (1 meter). Although growth tends to be similar across all varieties, some, like cherry tomatoes, are prolific, while others tend to be more frail.

To distinguish between the many tomato varieties, you can examine various criteria related to each plant's morphology, growth habit, and fruit.

Growth Habit

Most cultivated tomatoes are indeterminate varieties, meaning that, like wild tomatoes, they grow as vines with a seemingly indefinite lifespan. Given the right conditions, the plants will grow continuously until their natural decline. These plants therefore need to be trained and pruned. To identify indeterminate seedlings, look for thick stems that become increasingly thin as the plant grows and ages.

In temperate regions, you'll also find determinate varieties. The gene behind this mutation, called "SP" for "self-pruning," first occurred spontaneously in 1914, in Florida. Once determinate plants no longer produce flowers, they stop growing altogether. At this point, the plant stems end with a flower cluster at the very top. Determinate tomatoes are more bushlike, tend to reach maturity early, and produce smaller fruits, ranging from cocktail to medium-sized tomatoes. These varieties do not need to be pruned and trained, which makes them ideal for places like small gardens and balconies or patios, where growing space is limited.

Among the determinate varieties that can easily be grown in containers, you'll find micro dwarf tomatoes, which range in height from about 8 to 20 inches (20–50 cm), at most. They mature early and produce small fruits.

Some varieties are also referred to as "semi-determinate." These plants behave like determinates for the time it takes to set two or three fruit clusters, then they will start to grow again. With semi-determinates, you can therefore extend your tomato harvest.

Fruits

The tomato fruit is made up of flesh and seeds and is classified as a berry. Its appearance may change significantly from one variety to the next, but the most common shape is round, slightly flattened and ribbed. Tomatoes also contain hollow cavities, more or less depending on the variety, and vary in size. Although there is no rigorous system to classify them by size, here is an outline of a few broad categories:

- Less than 0.4 oz. (10 grams): currant tomatoes, which generally are not favored for eating
- From 0.4 oz. to 0.7 oz. (10–20 g): cherry tomatoes, the best-defined category

- From 0.7 oz. to 1.4 oz. (20–40 g): cocktail tomatoes
- From 1.4 oz. to 3.5 oz. (40–100 g): medium-sized tomatoes
- From 3.5 oz. to 10.5 oz. (100–300 g): large-sized tomatoes
- Over 10.5 oz. (300 g): beefsteak tomatoes

Color

You can generally guess a tomato's flavor based, in part, on its color. The skin and flesh both contain antioxidants and pigments. In the flesh, the main pigments are lycopene, beta-carotene, delta-carotene, chlorophyll, phytoene, and phytofluene. Lycopene is the pigment that makes tomatoes turn red. The skin contains a pigment that can either be clear, allowing the color of the flesh to show through, or yellow, which adds a layer of pigment over the color of the flesh. Red is the most common color, found in two thirds of tomato varieties.

Each tomato's acidity depends on combined acidic compounds in the fruit, like caffeic acid and ascorbic acid. It will also vary according to exposure to sun. When tomatoes are grown outdoors, they are naturally more acidic. Sugar content increases as the fruit ripens. Then, towards the end of the season, tomato acidity decreases.

Thanks to chemical compounds like vitamin C (especially found in raw tomatoes), lycopene (especially found in cooked tomatoes), carotenes (highly concentrated in orange varieties), and tomatine (found in green tomatoes), tomatoes contribute to preventing cancer and cardiovascular disease. However, when grown in excessively hot (above 95°F, 35°C) conditions, they lose some of their nutritional value.

Maturity

The term "maturity" refers to the number of days a tomato plant needs to produce mature fruit. The countdown starts the day you transplant tomato seedlings into your garden bed or container, and not from the seeding date.

There are myriad tomato varieties, and each has its own particular strengths and nutritional benefits that make it worth growing. Over the years, a few varieties have stood out in Jean-Martin Fortier's books for their vigorous growth in the field or under cover, for their high yields, and for their flavor. Some of these have been grown on his Quebec farm, Les Jardins de la Grelinette, since day one. Others have been grown by the many microfarms currently applying the Market Gardener Method in Europe and North America. All of them are easy to obtain from well-known seed suppliers. Some varieties are hybrids and crossbred plants that provide specific advantages, such as better disease resistance and particularly high yields, without compromising on taste or flavor. So many criteria are important to Jean-Martin Fortier when it comes to tomatoes, which he still considers to be the most emblematic produce at his farm stands.

Jean-Martin Fortier's 20 Favorite Tomatoes

Jean-Martin Fortier's 20 Favorite Tomatoes

Apéro F1

In French, *apéro* refers to drinks had before dinner or lunch. Thus, as the name suggests, Apéro F1 is a hybrid cherry tomato that is ideal for appetizers, aperitifs, or cocktails.

Characteristics
ORIGIN: France
COLOR: Red
SIZE: Cherry (0.6–0.7 oz., 18–20 g)
MATURITY: Early
GROWTH HABIT: Indeterminate
SHAPE: Small, oval
AVERAGE PLANT HEIGHT: 4 ft., 120 cm

Description
Highly productive, with uniform fruits. Consistent, long fishbone trusses.

A Note from Jean-Martin Fortier
Highly disease resistant. Rich flavor, very thin skin. Sweet and delicate taste.

Jean-Martin Fortier's 20 Favorite Tomatoes

Auriga

This variety was first distributed by Dr. Martin Stein in the former East Germany and later rediscovered by the Swiss after the fall of the Berlin Wall.

Characteristics
ORIGIN: Germany
COLOR: Orange
SIZE: Medium (2-3.5 oz., 60-100 g)
MATURITY: Mid-season

GROWTH HABIT: Indeterminate
SHAPE: Medium, round, slightly oval
AVERAGE PLANT HEIGHT: 6.5 ft., 200 cm

Description
Very hardy. Consistent, high-quality yields starting mid-season. Large quantity of trusses with 4 to 7 tomatoes each. Smooth skin. High carotene content.

A Note from Jean-Martin Fortier
Vigorous and hardy. Suitable for cold regions. Not prone to cracking. Firm and juicy, with a thick skin. Rich, sweet flavor. Use in salads and sauces.

Jean-Martin Fortier's 20 Favorite Tomatoes

Beefsteak

Beefsteak is an American tomato similar to Coeur de Boeuf (Cuor di bue), which has been bred into many other varieties.

Characteristics
ORIGIN: United States
COLOR: Red
SIZE: Large to beefsteak (7-18 oz., 200-500 g)
MATURITY: Late
GROWTH HABIT: Indeterminate
SHAPE: Ribbed, flattened
AVERAGE PLANT HEIGHT: 6.5 ft., 200 cm

Description
Very large fruit. When sliced, the tomato flesh looks meaty. Contains few seeds.

A Note from Jean-Martin Fortier
With a meaty, melt-in-your-mouth texture, this tomato has a sweet, complex flavor. Use in salads or as stuffed tomatoes.

Jean-Martin Fortier's 20 Favorite Tomatoes

Black Cherry

Known to be one of the best cherry tomato varieties, Black Cherry was bred by Vince Sapp, for the Tomato Growers' Supply.

Characteristics
ORIGIN: United States
COLOR: Purple to black
SIZE: Cherry to cocktail (0.5–1 oz., 15–30 g)

MATURITY: Mid-season
GROWTH HABIT: Indeterminate
SHAPE: Round
AVERAGE PLANT HEIGHT: 5.5 ft., 170 cm

Description
Productive until late summer or early fall. 10 to 12 fruits per cluster. Brown flesh. Harvested while still green, ripens well after harvest.

A Note from Jean-Martin Fortier
Well adapted to warm regions, cool climates, and short seasons. Consistent bushy growth. No pruning required. Fairly disease resistant. Not prone to cracking. Juicy, sweet, and tangy. Rich and complex flavor.

Jean-Martin Fortier's 20 Favorite Tomatoes

Black from Tula

This variety originated in Tula and was mentioned by Carolyn J. Male in her book, *100 Heirloom Tomatoes for the American Garden*.

Characteristics
ORIGIN: Russia
COLOR: Red to brown
SIZE: Large (9 oz., 250 g)
MATURITY: Mid-season

GROWTH HABIT: Indeterminate
SHAPE: Somewhat flattened and ribbed
AVERAGE PLANT HEIGHT: 7 ft., 220 cm

Description
High-yielding. Green shoulders, dark red flesh. Slight ribbing at the shoulders.

A Note from Jean-Martin Fortier
Good mid-season yields. Sweet and highly flavorful with meaty flesh. Use in salads.

Jean-Martin Fortier's 20 Favorite Tomatoes

Black Krim

This heirloom variety from the Black Sea region of Crimea is extremely popular with tomato lovers.

Characteristics
ORIGIN: Ukraine
COLOR: Purple, turns to mahogany once mature
SIZE: Large to beefsteak (4–5 oz., 120–150 g, and up to 18 oz., 500 g)
MATURITY: Mid-to-late season
GROWTH HABIT: Indeterminate
SHAPE: Round, slightly flattened
AVERAGE PLANT HEIGHT: 5 ft., 150 cm

Description
Prolific producer. Dense flesh. Shoulders are darker and often green. Contains few seeds.

A Note from Jean-Martin Fortier
Good drought resistance, is not prone to disease. Does not tolerate excessive watering (splitting). Sweet, delicate flavor, nonacidic. Use in salads.

Bosque Blue

Lee Goodwin developed this variety from a cross between the University of Oregon's Blue and Amy Sugar Gem.

Characteristics
ORIGIN: United States
COLOR: Red, turning very dark blue at maturity
SIZE: Medium (1.4–2 oz., 40–60 g)
MATURITY: Mid-season
GROWTH HABIT: Indeterminate
SHAPE: Round
AVERAGE PLANT HEIGHT: 6.5 ft., 200 cm

Description
Highly productive. Anthocyanin content similar to blueberries. Dark red flesh.

A Note from Jean-Martin Fortier
Firm and refreshing fruit with a hint of vanilla. Use in salads.

Jean-Martin Fortier's 20 Favorite Tomatoes

Brandywine

A tomato plant that delivers irregularly shaped fruits, Brandywine has frequently been used in crosses—both intentional (Purple Brandy) and accidental (Lucky Cross).

Characteristics
ORIGIN: United States
COLOR: Pinkish-red
SIZE: Large (5–7 oz., 150–200 g) to beefsteak (up to 18 oz., 500 g)

MATURITY: Mid-to-late season
GROWTH HABIT: Indeterminate
SHAPE: Round, slightly flattened, and ribbed
AVERAGE PLANT HEIGHT: 6.5 ft., 200 cm

Description
Ribbing at the shoulders of the fruit. Productive. Harvest is more abundant in the second half of the season.

A Note from Jean-Martin Fortier
For healthy growth and to make sure your plants don't become leggy (long, thin, and pale), make sure they are exposed to plenty of light and sun. Juicy and dense flesh, remarkable flavor, vine-ripened taste. Versatile use in the kitchen: raw, in salads, cooked, in sauces, and in coulis.

Jean-Martin Fortier's 20 Favorite Tomatoes

Clementine

Clementine is another appetizer classic that combines good flavor and color!

Characteristics
ORIGIN: Ireland
COLOR: Yellow
SIZE: Gooseberry (0.2–0.3 oz., 6–8 g)
MATURITY: Early

GROWTH HABIT: Determinate
SHAPE: Small, very consistent
AVERAGE PLANT HEIGHT: 6.5 ft., 200 cm

Description
Fruit the size of a gooseberry or small round plum. Highly productive. About 100 tomatoes per plant, uniform in size.

A Note from Jean-Martin Fortier
Adapts well to cooler climates. No pruning required. Juicy, sweet, and very flavorful.

Jean-Martin Fortier's 20 Favorite Tomatoes

Coeur de Boeuf

The real Coeur de Boeuf (meaning "ox heart" in English) is called *Cuor di bue* in Italian and, as its name suggests, it is heart shaped. It has often been imitated, but never equaled! Fruits have a classic tomato flavor.

Characteristics
ORIGIN: Italy
COLOR: Red
SIZE: Large to beefsteak (10–21 oz., 300–600 g, up to 35 oz., 1 kg)

MATURITY: Early
GROWTH HABIT: Indeterminate
SHAPE: Very clear heart shaped
AVERAGE PLANT HEIGHT: 6.5 ft., 200 cm

Description
Productive. Contains few seeds. Coeur de Boeuf has been used to breed many different varieties, resulting in several colors, like red, orange, and white. Tomato flesh is dense and aromatic. These new varieties stand out for being highly productive and delivering higher yields than the original variety.

A Note from Jean-Martin Fortier
Prune misshapen and excess fruits to promote the development of a few exceptional specimens. Very sweet, with dense flesh and rich flavor. Use in carpaccio and salads.

Jean-Martin Fortier's 20 Favorite Tomatoes

Dorothy's Green

In the 1980s, Dorothy Beiswenger introduced this American variety to tomato collector Craig LeHoullier.

Characteristics
ORIGIN: United States
COLOR: Green
SIZE: Large to beefsteak (12–17.5 oz., 350–500 g)
MATURITY: Mid-season

GROWTH HABIT: Indeterminate
SHAPE: Ribbed and slightly flattened
AVERAGE PLANT HEIGHT: 5.5 ft., 170 cm

Description
Productive. Green shoulders. Develops a pink blush at maturity. Moderate foliage. Contains lots of seeds and mucilage, and multiple cavities (locules).

A Note from Jean-Martin Fortier
Slow growth at the start of season. Yields improve in the second half of the season. Fruits keep well after harvest. Juicy, dense flesh, mild, sweet, and spicy. Typical green tomato flavor. Use in salads or coulis.

Jean-Martin Fortier's 20 Favorite Tomatoes

Gregori's Altai

This variety comes from the Altai Mountains, on the border between Russia and China.

Characteristics
ORIGIN: Siberia
COLOR: Deep pink
SIZE: Large to beefsteak (7–17.5 oz., 200–500 g)

MATURITY: Early to mid-season
GROWTH HABIT: Indeterminate
SHAPE: Round, slightly flattened
AVERAGE PLANT HEIGHT: 4 ft., 120 cm

Description
6 to 7 fruits per cluster. Highly productive.

A Note from Jean-Martin Fortier
Well adapted to short seasons and warm climates. Fairly sensitive to late blight. Sweet and popular flavor, very dense flesh, ideal for carpaccio. Eat raw or cooked; use in salads, sauces, coulis, and as stuffed tomatoes.

Jean-Martin Fortier's 20 Favorite Tomatoes

Hawaiian Pineapple

Though the origin of this tomato is unknown, it was first sold commercially in the 1970s in southern Indiana, USA.

Characteristics
ORIGIN: Unknown
COLOR: Bicolor, orange and red
SIZE: Large to beefsteak (up to 2.2 lbs., 1 kg)
MATURITY: Late
GROWTH HABIT: Indeterminate
SHAPE: Round, wide, ribbed, and slightly flattened
AVERAGE PLANT HEIGHT: 6 ft., 180 cm

Description
Highly productive. Irregular shape. Contains few seeds.

A Note from Jean-Martin Fortier
Very fruity and sweet, slight exotic taste. Smells like a pineapple when ripe. Mostly eaten raw, in salads and carpaccio.

Jean-Martin Fortier's 20 Favorite Tomatoes

Jaune de Saint-Vincent

This heirloom variety from France is one of the best yellow tomatoes.

Characteristics
ORIGIN: France
COLOR: Yellow
SIZE: Large (5.5–9 oz., 150–250 g)
MATURITY: Mid-season

GROWTH HABIT: Indeterminate
SHAPE: Round, slight ribbing
AVERAGE PLANT HEIGHT: 6.5 ft., 200 cm

Description
Highly productive. Smooth skin, distinct ribbing at the shoulders.

A Note from Jean-Martin Fortier
Well adapted to short seasons and cold regions. Juicy, sweet, mild, and tangy at the same time. Quickly becomes mealy once ripe. Use sliced or in salads.

Jean-Martin Fortier's 20 Favorite Tomatoes

Liguria

This variety is native to Italy and is sometimes referred to as Ligurian Coeur de Boeuf.

Characteristics
ORIGIN: Italy
COLOR: Red
SIZE: Large to beefsteak (7 oz., to 2.2 lbs., 200 g to 1 kg)
MATURITY: Early
GROWTH HABIT: Indeterminate
SHAPE: Ribbed pear
AVERAGE PLANT HEIGHT: 6.5 ft., 200 cm

Description
Productive. Pear-shaped fruit. 3 to 5 tomatoes per cluster. Contains few seeds.

A Note from Jean-Martin Fortier
Fairly disease resistant. Dense and pulpy fruit. Use in coulis and sauces.

Jean-Martin Fortier's 20 Favorite Tomatoes

Marmande

Developed by Vilmorin & Cie in the second half of the 19th century, this variety is a showstopper in vegetable gardens.

Characteristics
ORIGIN: France
COLOR: Red
SIZE: Large (5–9 oz., 150–250 g)

MATURITY: Early
GROWTH HABIT: Semi-determinate
SHAPE: Round, slight ribbing
AVERAGE PLANT HEIGHT: 6.5 ft., 200 cm

Description
Vigorous, irregular-shaped fruits, highly productive. 4 to 5 tomatoes per fruit cluster.

A Note from Jean-Martin Fortier
Will adapt to cold climates and short seasons, has good disease resistance. Typical heirloom flavor, complex, with high sugar content. Use in sauces, in salads, or stuff and cook au gratin.

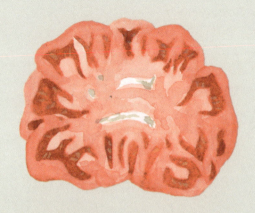

Marnero F1

Marnero F1 is a hybrid tomato from Marmande that was created for improved disease resistance.

Characteristics
ORIGIN: France
COLOR: Crimson, turns almost black once mature
SIZE: Large (7–9 oz., 190–250 g)
MATURITY: Early
GROWTH HABIT: Indeterminate
SHAPE: Ribbed
AVERAGE PLANT HEIGHT: 5 ft., 150 cm

Description
Highly productive. Uniform fruits. Dense flesh. Shoulders are darker and often green. Contains very few seeds.

A Note from Jean-Martin Fortier
Good pest and disease resistance. Once a third fruit has set, prune the remaining flowers on the truss. Sweet, with complex flavor. Use in salads, for cooking, and as stuffed tomatoes.

Jean-Martin Fortier's 20 Favorite Tomatoes

Orange Queen

When it comes to orange tomato varieties, Orange Queen is a favorite among gardeners for its unique and surprising color.

Characteristics
ORIGIN: United States
COLOR: Orange
SIZE: Large (4–6 oz., 120–180 g)

MATURITY: Early
GROWTH HABIT: Indeterminate
SHAPE: Round, lightly ribbed shoulders
AVERAGE PLANT HEIGHT: 5 ft., 160 cm

Description
Contains few seeds. 4 to 5 tomatoes per cluster.

A Note from Jean-Martin Fortier
Juicy, firm, and dense tomato. Sweet and mild taste. Use in salads, in coulis, and as stuffed tomatoes.

Jean-Martin Fortier's 20 Favorite Tomatoes

Pineapple

This heirloom variety was revived in the 1950s and remains one of the best from a flavor standpoint.

Characteristics
ORIGIN: United States
COLOR: Bicolor, yellow-orange streaks
SIZE: Large to beefsteak (9-14 oz., 250-400 g, up to 2 lb., 1 kg)
MATURITY: Late
GROWTH HABIT: Indeterminate
SHAPE: Slightly flattened, ribbed
AVERAGE PLANT HEIGHT: 6.5 ft., 200 cm

Description
Good yields. Skin and flesh are the same color. Shoulders are quite wide and sometimes not mature. Firm and dense with very few seeds.

A Note from Jean-Martin Fortier
Better suited to warm and sunny regions. Sweet and complex flavor, not very juicy. Use in salads and carpaccio. Loses some flavor when cooked.

Jean-Martin Fortier's 20 Favorite Tomatoes

Rose de Berne

This Swiss heirloom variety is quite famous in France and is considered one of the best tomatoes because of its flavor.

Characteristics
ORIGIN: Switzerland
COLOR: Light pink
SIZE: Large (4–6 oz., 120–180 g)

MATURITY: Mid-season
GROWTH HABIT: Indeterminate
SHAPE: Round, slightly flattened
AVERAGE PLANT HEIGHT: 6.5 ft., 200 cm

Description
4 to 6 tomatoes per cluster. Average yields.

A Note from Jean-Martin Fortier
Late blight resistant. Prone to cracking. Sweet and rich flavor, with dense, juicy flesh. Use in salads.

For professional market gardeners and home gardeners alike, growing your own vegetable seedlings is not only a source of joy and pride but also the best way to be in control of your garden planning and to master every step in the market gardening process.

To begin, sow seeds, then pot up the seedlings; a few weeks later, transplant the seedlings into the ground, either in a tunnel or greenhouse or in the field. For microfarms and home gardens, growing plants from seed is the most common way to propagate vegetables. Tomatoes are no exception. The process can sometimes be tricky and requires patience, precision, and sustained attention for about 6 to 8 weeks. As soon as the seeds germinate, the seedlings start growing and require constant daily monitoring and care—without fail—until they are potted up.

Growing Seedlings

Seeding

Tomatoes should be seeded about 6 to 8 weeks before the last frost date in your climate.

Seedlings need consistent temperatures, between 68°F and 77°F (20°–25°C), and soil that is damp but not waterlogged. They can be started in a greenhouse, in a cold frame or mini greenhouse equipped with heating mats, or in a bright, heated room in your home. Seeds usually germinate in about 5 to 7 days and up to 14 days with some varieties.

After seeding, dip the flat or containers in water to fully wet the soil. You can also use a watering can with a fine spray head, a mister, or even a normal water bottle with small holes poked into the cap.

Once the seeds have germinated, cotyledons appear and start the photosynthesis process. From then on, your tomato plants will need lots of light, about 10 to 12 hours a day.

Seeding into Pots

Best suited to small-scale growers
with no more than a few dozen seedlings,
this method is primarily used by home gardeners.

Purpose and Advantages

Starting seedlings in pots is easy and provides one significant advantage: you avoid the delicate and time-consuming potting-up stage. You can use pots made from PVC or from compressed peat, which is cheaper and more eco-friendly. Since it breaks down in the soil, you can transplant the pot directly into the ground, without having to remove the plant.

Doing It Right

1 Before seeding, fill your containers with potting mix. You can also add 5% perlite, to help improve drainage. Water the soil and wait for it to dry, going from being wet to moist.

2 Drop the seeds onto the surface, then lightly cover them with soil and spread evenly. Lightly tamp the surface, to ensure good contact between the seeds and the soil, but be careful not to overdo it. If the soil becomes too compact, the seeds will struggle to sprout.

3 Water with a gentle spray or mist, and label each variety you seeded. Put the pots in a bright and warm space, water sparingly until germination, then water more regularly as the seedlings develop.

Seeds

Tomato seeds are quite small. A single ounce can contain thousands of seeds. You can either save seeds yourself (see p. 48) or buy them from a producer. Commercial seeds come in different forms: bare seeds, which have been dried and freed of plant debris; seeds coated in clay, which makes them slightly larger, more uniform in size, and thus easier to work with a hand seeder; seed tape, sheets, or mats, where the seeds sit between thin biodegradable pieces of paper that can be applied directly onto soil without having to calculate spacing.

On the seed packet, you'll find a description of the plant and growing instructions.

All seeds have a lifespan after which germination rates start to decline significantly. For tomato seeds stored under the right conditions, this is roughly 5 years.

Tip from Jean-Martin Fortier

With tomato seedlings, the main threat is damping-off, a disease caused by various fungi (*Fusarium*, *Pythium*, *Phytophthora*...). It is due to excessive humidity combined with either a lack of light or temperatures that are too high (above 82°F, 28°C), or both. As soon as condensation forms on the walls of your mini greenhouse, cold frame, or greenhouse, you must increase airflow to control humidity levels. To further reduce the risk of damping-off in seedlings, you can incorporate powdered charcoal into your potting mix before seeding.

Seeding into Open Flats

Open flats, aka flats, seed flats, or seeding trays, are wide plastic containers with high edges. You can use them to sow large quantities of seeds.

Purposes and Advantages

Seeding into flats is a method used by both home gardeners and professional market gardeners. Flats sold commercially come perforated and in standard sizes. Some mini heated greenhouses even come with integrated seed flats. Home gardeners can use any sturdy plastic or polystyrene container, as long as it is flat, with a sufficient surface area and fairly high edges. Note that repurposed polystyrene containers provide two advantages: they are insulating and eco-friendly, since you are giving them a second life. If your container is not already perforated, drill several holes into the bottom so water can drain and not saturate the soil.

Tip from Jean-Martin Fortier

When seeding into a flat, be careful not to sow too many seeds as you may end up with seedlings that are weaker and likely to become leggy. Instead of developing normally, they grow tall and turn a pale yellow. This can also result from other factors like a lack of light and a nutrient deficiency.

Growing Seedlings

Doing It Right

1 Drill or poke holes into the bottom of the container you selected, if it is not already perforated. Add a layer of soil. To sow the seeds, fold over one side of the packet to channel their distribution onto the soil.

2 Cover seeds with a thin layer of sifted soil, using something like a strainer or screen for Mason jars.

3 Tamp the soil, with a wooden slat, to increase contact between seeds and soil, then water with a gentle spray or mist.

4 Label and transfer into a greenhouse, in a bright and warm space. Water sparingly until the seeds germinate, then more regularly throughout the seedling stage.

Seeding in Plug Flats

Seeding into plug flats is a useful method for growing many seedlings. Precise and efficient, it is most suitable for professional market gardeners.

Purpose and Advantages

To promote good root development in your tomato seedlings, select flats with roughly 70 cells. This approach optimizes time spent starting seedlings but also speeds up potting. It's much easier to remove them from their cells and transplant the entire soil plug and root ball without exposing the roots.

To make sure your tomato seeds germinate well, opt for a very high-quality soil. Place the trays on heating mats to maintain a steady soil temperature, promoting good germination rates and healthy seedling development.

By sticking to these principles, you can grow seedlings that will be ready to be potted up in just 30 days.

Doing It Right

1 Fill the cells with potting mix, then level the soil by hand or with a wooden slat. Set the trays in water that is about half as deep as the tray, soaking them to moisten the soil. Next, put them in the greenhouse and let the soil drain.

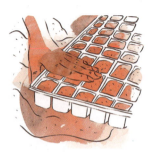

2 Drop one seed into each cell, using tweezers or a seeder. Cover all seeds with a thin layer of soil and tamp lightly to improve contact between the soil and the seeds.

3 Water and label the trays, then put them on heating mats to maintain soil temperatures at 75°F (24°C) until the seeds germinate. You can also put a dome over the plants to keep the soil moist. After the seeds sprout, remove the heating mats and maintain air temperatures around 75°F (24°C) during the day and at least 65°F (18°C) overnight.

Tip from Jean-Martin Fortier

To prepare for outdoor living conditions, seedlings go through a hardening-off process, after growers transfer them to an acclimation zone where temperatures are cooler. In this stage, you can add a small pinch of nitrogen-rich chicken manure to each cell.

Grafted Tomatoes

Grafting is an ancient practice that involves splicing together two varieties. For the base, you want to select a rootstock that is disease resistant and that will develop a broad network of roots; this is often a hybrid variety like Alligator or Arnold. The top part, called a "scion," should be selected for its vigorous growth and quality fruit.

About one month after the seeding date, cut the rootstock roughly 1.25 to 2 inches (3–5 cm) above the soil surface, using a very sharp disinfected knife. Then, with the same knife, separate a tomato seedling (scion) from its base. Take the scion and line it up with the cleft you've made in the rootstock. Tie these parts together with raffia twine to allow the plant tissue to heal.

When planting your grafted seedlings, make sure to keep the graft union above the soil line in open air.

Grafted plants are stronger and more disease resistant. They also deliver better yields. Consider grafting if you have unfavorable soil (wet, high-clay content).

It is worth noting that grafting requires an additional investment, both in terms of materials—seeds, rootstock, and potting mix—and in terms of labor. This approach will not necessarily be profitable if your tomato growing season is short. Sometimes, a simple crop rotation is enough to prevent the spread of disease from one year to the next.

Saving Your Own Seeds

To grow tomatoes without having to buy new seeds every year, you can collect seeds from a portion of your own harvest.

CHOOSING THE RIGHT SEEDS
Pick the biggest seeds from a ripe fruit of a healthy plant, ideally from one of the first tomatoes of the season.

As a precaution, we recommend saving more seeds than necessary, in case you encounter germination issues. To check germination rates, you can use a fairly simple test: when submerged in water, viable seeds should sink. You can then discard any floating seeds.

EXTRACTING SEEDS
Harvest a ripe tomato, cut it in half horizontally and squeeze it over a glass or cardboard cup. Each container should hold only one variety and be labeled with the harvest date and variety name. Once the extracted seeds are covered in liquid from the tomato, they will begin the fermentation process. With some varieties, you may want to add a little water.

FERMENTATION
Next, store the containers in a well-ventilated and temperature-controlled space kept above 68°F (20°C). Fermentation usually takes 2 to 3 days, perhaps less depending on conditions. At the end of this process, the surface will have developed a thin layer of mold, caused by the fungus *Geotrichum candidum*. It destroys mucilage (the gel that envelops the seed and prevents germination while it is still in the fruit) and microbes that might cause bacterial diseases.

Be careful, white tomato varieties develop less mold than others.

RINSING
After pouring the seeds into a fine-mesh sieve and removing the layer of mold, rinse them thoroughly under running water while moving them around. Rinse off as much mucilage as you can. If some mold is left over, it's not a big deal. To get rid of as much residual water as possible, run a sponge along the underside of the sieve.

DRYING
To dry them, transfer any wet seeds into coffee filters. Never use plastic bags in this step. On the filters, write the variety name and harvest date, then hang them up in a dry and well-ventilated room for at least 10 days, away from direct sunlight. Once the seeds are completely dry, it should be easy to pull them off the filters. If you are only saving a small number of seeds, dry them on a piece of white cloth or paper towel.

STORAGE
Dried seeds can be stored in coffee filters, envelopes, or plastic bags. Their useable lifespan will be about 5 years; after that, the seed viability will diminish over time.

To store them indefinitely, you can freeze the seeds in airtight containers.

Potting Up

About 3 to 4 weeks after sowing, tomato seedlings should each have two real leaves that are 4 to 6 inches (10–15 cm) long. If they are in an open flat or plug flat, the seedlings will start to run out of space. At this stage, it is still too early to transplant them into your garden beds because they are young and fragile. The next, intermediate step—potting up—will give them the time and space to keep growing and to become strong enough to withstand outdoor living conditions. In this step, each seedling is transferred from the seed flat or plug flat into a larger pot, about 4 inches (10 cm), filled with more high-quality potting mix amended with compost.

The potting-up process is easy but requires a delicate touch to avoid damaging the roots. Only select healthy seedlings; frail and diseased seedlings should be set aside and can be composted.

Potting Up:
From Open Flats to Pots

In the potting-up stage, seedlings grown in flats are transferred to smaller containers, so that each can have its own space to grow. This helps seedlings prepare to be transplanted into the ground.

Purpose and Advantages

The process of potting up tomatoes from open flats to pots is simple and effective. This approach is primarily used by home gardeners, and sometimes by professional market gardeners.

To pot up tomatoes, growers use individual containers made from a flexible material. Some are made of biodegradable peat, which can be planted right into the ground, without extracting the seedling. The downside is you must replace these containers every year.

Plastic pots, however, can be reused for several years if you take care of them. For both methods, you first need to figure out how many pots you need.

If you sow tomatoes in an open flat, you can either pull out a clump of seedlings, then separate and pot them up individually, or extract them one by one.

Doing It Right

1 Using a thin stick, or simply your fingers, lift and remove one or more seedlings from the tray. Work carefully to avoid damaging the roots.

2 Fill your pot with fresh soil and, using a stick, dig a hole that is wider than the root ball you are about to plant, leaving 1 or 2 inches of soil in the bottom of the pot.

Potting Up: From Open Flats to Pots

3 Without forcing, gently put the seedling in the hole and fill it up to the cotyledons. You can use a wooden stick to guide the stem and roots into the hole. Lightly tamp the soil without compacting it to improve contact with the roots.

4 In one crate, place seedlings of the same variety and label it. Water with a gentle spray, then put the seedlings in a greenhouse, in a cold frame, or under covered hoops.

Tip from Jean-Martin Fortier

If the root ball stays together when pulled from its cell, the plant is ready to be potted up. By this time, the seedling will have grown a fair bit and consumed all available nutrients in the soil. It now needs a boost to keep growing. Potting up the seedling gives it extra space and, especially, fresh high-quality soil mixed with compost to prepare it for transplanting in the field or garden. But before that happens, it must spend another 3 to 4 weeks in a room heated to between 64°F and 68°F (18°–20°C).

Potting Up:
From Plug Flats to Pots

Seedlings grown in plug flats will have a larger root ball than those grown in open flats. This means they can be potted up sooner as they are already separated and growing independently.

Purpose and Advantages

Professional market gardeners typically use plug flats to grow their seedlings. This method is efficient and effective as it makes it easier to monitor and care for seedlings and limits handling time, since they are in one tray. But just like those growing in open flats, they will eventually need fresh, nutrient-rich soil and extra space for root development.

Before potting up seedlings, it's best to water them and fully wet the root ball so that they will be easier to pull from their cells. You can even soak them in a fertilizer solution (beet juice or chicken manure diluted in water) that will strengthen the seedlings, make them easier to extract, and help them better tolerate the stress of the potting up and subsequent transplanting processes.

Tip from Jean-Martin Fortier

To prepare seedlings for the shift to outdoor conditions, we recommend an intermediate stage: hardening off. A week before the transplanting date, add a little chicken manure and begin to either lower the temperature in greenhouses and shelters, especially at night, or reduce the water supply if you cannot lower the ambient temperature. Plants have to experience at least 4 to 5 days of lower temperatures. It is essential that they continue to receive sufficient direct light, otherwise they may become leggy. If you expect a late frost, radiation frost, or extreme cold, bring the seedlings inside temporarily for protection.

Doing It Right

1 To pull a seedling from the tray, gently hold it by the stem while lightly squeezing the sides and bottom of the cell.

2 Fill a pot approximately 1/3 of the way up with soil, place the seedling in the center, then top up the soil, making sure to bury part of the tomato stem.

3 Lightly tamp the soil with your hands to ensure good contact between the roots and the substrate. Water with a gentle spray, label each seedling with the variety name and put all plants in a greenhouse, in a cold frame, under covered hoops, or in a bright, heated room.

The tomato planting process entails more than just putting a seedling in the ground. It involves a set of operations, starting with careful soil preparation, including loosening the soil and leveling the bed surface. Next comes the actual planting, when the tomato seedlings go in the ground. Professionals plant tomatoes in beds, in field blocks, or under shelter, while home gardeners use planter boxes or raised beds.

Before doing any tomato plant care, you must set up supports to train the seedlings, either right after planting or within a few days, but no later. Your trellising or staking system will depend on whether your tomatoes will grow under shelter or outdoors and whether the soil is mulched with an organic or synthetic ground cover.

Planting and Early Care

Preparing the Soil

For quality harvests, healthy soil is your best guarantee. Depending on its composition and structure (clay, light soil, rocky, etc.), bed preparation time will vary. That's why it's important to factor this work into your farm plan. If you have rocky soil, for instance, you'll probably need to set aside time to go through beds with a walk-behind tractor at least once, and well in advance of the growing season. However, if you regularly maintain your beds, a single pass with a wheel hoe and broadfork should be enough to loosen the soil, both in depth and at the surface, while also weeding. With these tools, you can avoid mixing and turning over different soil layers, which is better for soil health. Finally, in the winter preceding your tomato season, amend the soil with compost or manure, so that it will contain all the organic matter that tomatoes need to thrive. The goal is to end up with soil that is loose, highly fertile, and nutrient rich. This will allow your plants to develop and produce high-quality fruit.

Permanent Beds

For professionals, permanent beds are a highly effective method of cultivation, both in the field and under shelter. These slightly raised beds ensure high yields, but do require some preparation.

Purpose and Advantages

To grow bountiful field tomatoes, make sure that you are planting high-quality seedlings and that the soil has been properly prepared. This involves loosening, weeding, and amending the soil. Don't plant tomato seedlings until after the last frost and wait for the soil to reach at least 61°F (16°C). For tomato seedlings, ideal temperatures are around 64°F to 66°F (18°–19°C). A tip to warm up your soil more quickly: cover your beds with a transparent plastic sheet or a floating row cover.

Planting and Early Care

Doing It Right

1 Go through your beds with a broadfork to loosen the soil in depth, then amend them by applying a roughly 2-inch (5 cm) layer of compost.

2 Using a bed preparation rake, level the surface while spreading the compost and evenly incorporating it into the soil.

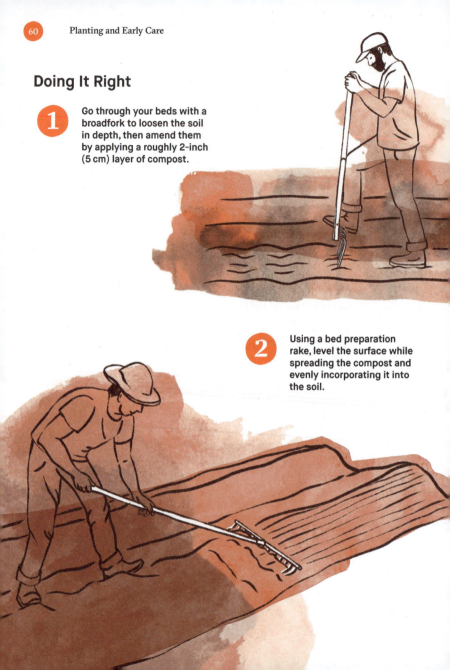

Permanent Beds

3 To promote good root development in loose soil, run a tilther down the bed (set to a 2-inch, 5 cm, depth), to create a finer soil texture and level the surface.

Tip from Jean-Martin Fortier

To avoid any unpleasant surprises, make sure there are no weed seeds or roots in your compost. When buying compost, you must select compost that contains no weed seeds, which could take over your beds. You can also make your own, using organic waste and even manure, if you are raising livestock.

Planter Boxes and Garden Mounds

Home gardeners growing just a few tomato plants may opt to use a planter box or a garden mound, also called a *hugelkultur* bed. Both require good soil preparation.

Planter Boxes

Planter boxes are ideal for gardening in small spaces. To make your own, build a square wooden frame with sides that are 3.5 to 4 feet (100–120 cm) long and 12 to 16 inches (30–40 cm) high, then fill the box with potting mix. At the corners, slot the boards into metal brackets to reinforce the frame. Make sure to cover the bottom and side walls with a geotextile fabric to protect the frame. For a more elegant look, you can decorate the outside of your planter box with a wooden trellis, reed fencing, or brushwood fencing. With a planter box, you can grow a higher density of vegetables and look after crops easily. Since the surface area is smaller, watering and weeding requires less time. Because the vegetables are more accessible in a raised bed, gardening work puts less strain on your back. If you're building multiple planter boxes side by side, plan to have fairly wide paths between them, so it is easy to get around.

Garden Mounds (*hugelkultur* beds)

If you are working with infertile, acidic, poorly drained, or low-quality soil, this mounding technique is an ideal solution. Start by loosening and hoeing the soil, then cover it with cardboard. Next, build a domed base from logs and sticks covered with branches, and layers of dead leaves, wood chips, straw, grass, and potting mix. One main advantage of garden mounds is that they last for several years. As the wood decomposes, it provides the organic matter that vegetables need, while keeping the soil cool and moist at all times. To further extend the lifespan of your garden mounds, build them even higher by setting up a 1-to-1.5-foot (30–50 cm) barrier on each side,

using stones, logs, or boards. Plus, the higher bed makes it more comfortable for gardeners to tend to the plants.

Fertilizing the Soil Before Planting

Tomatoes are heavy feeders, consuming a lot of nutrients, and they stay in the ground for a long time, at least 100 days. To best meet their needs, plan to significantly amend the soil when preparing it for the seedlings. In addition to compost, we recommend spreading a nitrogen-rich organic fertilizer, like pelleted chicken manure or alfalfa meal, a few handfuls per square yard or square meter. You may also choose to add potassium sulfate or a solution containing potassium and magnesium sulfates. These last inputs will improve your plants' disease and drought resistance and make them more cold hardy, while also enhancing their fruit's flavor.

Planting

Tomatoes should be planted between mid-April and mid-May, depending on the climate, always after the last frosts.

To develop a good fruit set, tomato plants need plenty of light, sunshine, and temperatures between 64°F and 82°F (18°–28°C). Below 55°F to 59 F (13°–15°C), they grow poorly and may develop misshapen fruit. Above 90°F (32°C), most varieties will encounter problems with pollination and fruit set.

If you want to harvest tomatoes by the end of June, growing them in a tunnel or greenhouse is highly effective. With this approach, you can start your seedlings at the end of February, and pot them up in early April.

Tomatoes grow best in light-textured soil that has a high organic-matter content. In terms of daily water requirements, tomato plants consume an average of 0.25 to 0.5 gallons (1–2 L). In sandy soil, you may need to water two or three times as much, whereas the water input can be halved for clay soils.

Planting in Greenhouses or Tunnels

Professionals are usually the only ones who grow vegetables under shelter as this requires a greenhouse or tunnel. However, some home gardeners may choose to use tomato tunnels to benefit from a sheltered growing space.

Purpose and Advantages

Planting under cover allows growers to increase crop density. You can grow tomatoes in two rows, or in double rows set 2.5 to 3.5 feet (80–100 cm) apart. Within the row, plant seedlings 2 feet (60 cm) apart, then stagger the plants in adjacent rows. This means shifting all the plants in one row by 1 foot (30 cm) from the end of the bed or row. With this crop density, the foliage will create significant plant cover, shading the soil and limiting water loss to evaporation.

When planting tomato seedlings, make sure each one looks healthy and has developed enough; ideally, plants should be 16 to 20 inches (40–50 cm) tall. The foliage should be a vibrant green, which is a sign of good health, and the root ball should be moist so that it will continue growing in the ground.

Note that we recommend pruning any flower clusters from seedlings, so that they can focus primarily on developing roots, stems, and leaves.

Doing It Right

1. Along each bed, start by installing four drip irrigation lines, inserting pins every 5 feet (1.5 m) to secure them, then lay down a plastic mulch the width of the bed. Use a mulch hole burner to make 5-inch (12 cm) holes in the cover.

2. Dig a relatively deep hole, remove the tomato seedlings from their containers, and plant them deep enough to bury the stem, leaving one foot (30 cm) above the soil surface. Fill in any gaps around the root ball with soil, making sure no air pockets are near the roots.

3. Level the soil and top it up, if needed, then gently tamp to anchor the root ball in the soil. Finish by generously watering the plants to ensure further settling of the soil and provide the water that they need to take root.

Tip from Jean-Martin Fortier

To improve the root development in your tomato seedlings after planting, tilt the root ball in order to bury the first 2 to 4 inches (5–10 cm) of stem. This stem part then generates new roots, improving the plant's vigor and growth, and later fruiting.

Planting in the Field

Depending on your climate, you may be able to plant tomatoes in the ground without a shelter. This approach is especially relevant for home gardeners who may not have a greenhouse or tunnel.

Purpose and Advantages

When planting tomatoes without a shelter, lower-density spacing is preferred. Opt for single rows, with plants set 30 to 40 inches (80–100 cm) apart, rather than double rows. Maintain about 25 inches (60 cm) between rows. Plant your tomatoes outdoors only after the last frosts. You should also keep an eye on overnight temperatures and provide protection when cooler weather is forecast. If temperatures drop below 55°F (13°C), cover each row with a floating row cover supported by tomato stakes, hoops, or metal frames. You can even go so far as to place individual protective covers over each plant. This solution is better suited to home gardeners with few plants.

Tip from Jean-Martin Fortier

If you don't want to invest in a tunnel, or if you grow relatively few tomatoes, individual plant covers are a very simple solution that you can set up to protect plants from pests and mitigate the effects of bad weather, like wind, hail, or cold. Covering plants at the start of the seasonal creates a warmer microclimate around them that stimulates growth. Tie the cover to the top of the tomato stake so that every evening and on days when the weather is inclement, you can drop this perforated plastic around the plant, letting it reach the ground. Once daytime temperatures rise above 59°F or 60°F (15°–16°C), the cover is no longer necessary.

Doing It Right

1 To create a straight row, place a temporary string line down the center of the bed and then dig holes for your seedlings 2 feet (60 cm) apart.

2 With a spade, loosen the soil in each hole to a depth of 1 foot (30 cm), and remove any rocks and weed roots. Add fully decomposed compost, sifted and amended with one or two handfuls of horn meal, to the soil in the holes.

3 Remove a tomato seedling from its container and place it in the hole, burying the base of the stem. Fill the hole with potting mix, if needed, then gently tamp it around the root ball and water thoroughly.

Growing Tomatoes in Containers

You can grow tomatoes in containers on patios, on balconies, or in courtyards, as long as the location is bright and sunny.

CONTAINERS
We recommend terra-cotta pots, or even plastic pots, and always make sure that the bottom is perforated. First put a 1-to-2-inch layer of gravel (3–5 cm) in the pot to ensure proper drainage. Then add a piece of felt or burlap before pouring in the potting mix.

Choose a container that suits the variety you want to grow: indeterminate tomatoes need a volume of at least 10 US gallons (35–40 L) and about 1.5 feet (40 cm) in all directions; for determinates, choose 5 to 10 US gallons (20–40 L) and 1 to 1.5 feet (30–40 cm) in all directions; and for micro dwarf varieties, use 1 to 2 US gallons (5 to 10 L) and a depth of 1 foot (30 cm). If you are using a large container, space seedlings at least 2 feet (60 cm) apart, or 18 inches (40 cm) apart if planted in staggered rows.

FERTILIZATION
Tomatoes grown in containers require a high-quality and nutrient-rich potting mix and more fertilizer inputs. However, the nutrients in the soil will quickly be depleted, so regular fertilization is essential. Every 15 days, before watering, spread a few handfuls of tomato-specific organic fertilizer, which is available at your local garden center. You can also dilute nettle and comfrey tea (5% to 10% of the mixture) in a watering can and apply. See p. 99 for fertilizer teas.

IRRIGATION
A tomato plant grown in a 5-gallon (20 L) container needs roughly 1.5 to 2 quarts (1.5–2 L) of water per day. In hot weather, we recommend watering every evening. To keep the soil cool and moist, cover the surface with a mulch of bark, leaves, or straw.

GROWING TOMATOES IN BAGS
This is a creative way to grow tomatoes. Plant seedlings directly into 50-to-70-quart (50 to 70 L) bags of horticultural-grade potting mix. You can then poke a few holes along the sides, to allow the soil to drain and to avoid waterlogging. At the top of the bag, preferably in the center, create two or three holes with a 4-to-6-inch (10–15 cm) diameter, in which you plant the tomatoes.

TRAINING

Tomato plants have soft, tender stems that, if left unsupported, tend to collapse under the weight of their fruits and run along the ground. While they naturally grow this way, it is detrimental to the development of healthy foliage and high-quality fruit. When in contact with the soil, stems and fruit are exposed to water and soil splashing and, especially, to disease. This is particularly true of late blight that develops when vegetation is too dense and lacks airflow.

That's why we highly recommend training tomato plants as soon as they are in the ground, by attaching the stems to sturdy supports. The plants can then grow taller, which also makes them easier to care for, whether you are pruning or harvesting. There are several methods for training tomatoes, depending on how they are grown.

Training Field and Garden Tomatoes

Stakes can be made from various materials, such as wood or metal, but they must be strong enough to support stems, leaves, and fruit, and remain weather resistant throughout the growing season.

Purpose and Advantages

Training tomatoes with wooden stakes or metal rods is primarily for home gardeners, as this approach has several drawbacks that make it impractical for professionals. It requires not only owning stakes, which are both heavy and cumbersome, but also storing them once the season is over.

If you choose this method, put the stakes in the ground when planting your seedlings, then immediately tie them to the stakes. As the plants grow, keep tying the top of the stem to the stake. Many materials can be used for stakes. If you want something sturdy, opt for rebar or wooden stakes, particularly chestnut or acacia, which is known to be durable and weather resistant. Another option are stainless steel tomato spiral stakes that support one stem at a time. However, these may need to be reinforced by a piece of wood, to stop them from tilting under the weight of the fruit.

When growing several tomato plants side by side, you can tie the tops of the stakes together like a teepee, or set them up as a column supporting four plants. You can also create other shapes, like a V or an inverted V, as you might do with pole beans.

Lastly, you can surround the plants with wire mesh and attach it to wooden stakes, creating a circle with a 24-to-32-inch (60–80 cm) diameter. This method is best suited to cherry tomatoes, which require little pruning and will grow freely within their cage.

Doing It Right

1 Drive each stake roughly 1 foot (30 cm) into the ground, before or right after planting. In the latter case, be careful not to damage the roots when you bury the stakes.

2 Using cotton string or twine, tie the base of the main stem to the stake, but not too tightly: you need to leave room for the stem, which will thicken as it grows along the stake.

3 As the tomato plant grows, loosely tie the stem to the stake every 10 to 12 inches (25–30 cm) to keep it upright and ensure good support once the fruit starts to become heavy.

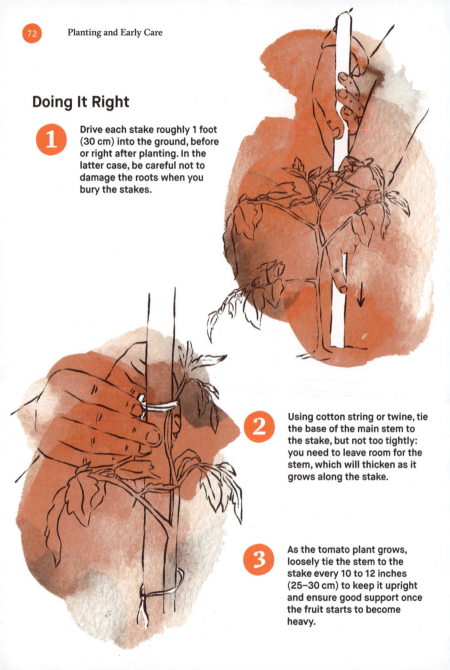

Tip from Jean-Martin Fortier

You can also opt to set up a sturdy string trellis. To do this, start by driving two chestnut stakes, each 5 feet (1.5 m) tall, into the ground at a depth of 16 inches (40 cm), with 4 or 5 tomato plants in between them.

Next, connect them horizontally, along the tops, using a wire or wooden slat. Tie strings along the top of the structure, so that each one hangs above a plant and is long enough to reach the ground. You can then tie the plants to the strings or wrap them around the stems as they grow.

It's important to ensure that the whole system is rigid and strong enough to support the weight of 4 or 5 tomato plants.

Trellising Under Shelter

In tunnels and greenhouses, professional market gardeners tend to use nylon line to trellis tomato plants. They are directly connected to the greenhouse or tunnel structure.

Trellising Under Shelter

Purpose and Advantages

With this trellising method, the setup is easier and quicker. It can be adapted to any type of greenhouse or tunnel, provided that the frame is strong enough.

Growers hang spools of nylon line from the shelter's metal structure, unwinding them until they reach the ground, to support the stems as they grow.

Doing It Right

 Loosely tie each nylon line to the base of the tomato plant, then gently wrap the line around the stem without damaging it.

 As the tomato grows, use plastic tomato clips to connect the stem and the string, leaving room for the stem to grow.

Tip from Jean-Martin Fortier

When using plastic tomato clips to connect the string to the stem, you only need them for the first 40 or 50 inches (100–120 cm). Beyond that point, you can simply wrap the string around the rest of the stem.

Mulching

Mulching protects the soil by covering it with a layer of organic matter, plastic sheeting, or woven synthetic fabric. This approach provides several advantages:

- It allows water and air to pass through the mulch, supporting plant development (unless it is plastic film or tarp).

- It prevents weed growth, which reduces time spent weeding.

- It limits soil water evaporation, so you do not need to water as often.

- It regulates soil temperature and reduces the root system's exposure to heat stress.

- It prevents soil leaching, stops soil and water from splashing onto the plants and fruits, and prevents soil runoff, especially for outdoor crops.

Organic Mulch

Organic mulches, which combine compost and straw, or any other untreated plant waste that does not contain weed seeds, protect and amend the soil.

Purpose and Advantages

Organic mulch is primarily used by home gardeners as it can be tedious to set up on a large scale. The mulch feeds microorganisms, improving soil fertility and texture. Be careful not to apply it too thick at the base of your tomato plants, to avoid suffocating the collar and to prevent diseases.

Doing It Right

1. Weed the soil between the plants, then spread 4 to 6 inches (10–15 cm) of plant-based mulch over the soil in an even layer.

2. When the crop is finished, bury the partially decomposed organic mulch to amend the soil.

Tip from Jean-Martin Fortier

Organic mulch can attract rodents that will devour your tomatoes, as well as slugs, which like moist environments. To avoid this issue, encourage birds to come near your crops to eat pests. While birds love fruit that is close to the ground, they rarely consume leaves or harm the plants. This is the advantage of having ecological niches, like hedges or flowering fields, on your farm or in your garden.

Synthetic Mulch

Geotextile felt, plastic films and tarps, and woven fabric are mainly used by professionals. This type of mulch requires good soil preparation.

Purpose and Advantages

Synthetic mulch prevents sunlight from reaching the soil, retains more moisture, and inhibits weed growth, but unlike organic mulch, it has no effect on soil fertility. It is highly practical and useful outdoors but is especially well suited to growing in greenhouses and tunnels. Before planting, lay the plastic over soil that has been loosened and leveled. Then make holes in it for your tomato seedlings. Although they are synthetic and therefore not very environmentally friendly, tarps are durable, reusable, and recyclable.

Doing It Right

1. Cultivate, fertilize, and level the soil, making sure to break apart any clumps.

2. Lay the tarp over the bed, eliminating any big gaps between it and the soil surface. Keep it in place with anchor pins set every 5 feet (1.5 m). In the field, you can leave the aisles uncovered, whereas it is common to cover the entire surface, including aisles, in greenhouses.

Tip from Jean-Martin Fortier

Two-toned plastic tarps are highly effective. The white side, facing up, reflects light at the tomato plants and distributes it throughout the greenhouse, while the black side, facing down, keeps the soil in darkness, retains moisture, and regulates soil temperature.

Companion Planting

By combining tomatoes and other vegetables, you can optimize your growing area, diversify production, and improve yields.

VEGETABLE CROPS

Tomato roots initially grow near the surface, so they can be combined with carrots, which have deeper roots. Carrots should be sown after the tomatoes are planted and will be harvested well before tomatoes begin to produce fruits. Lettuce is also an excellent neighbor for tomatoes. It doesn't grow invasively and will adapt to the tomatoes' regular watering schedule. Plus, it acts as a ground cover, limiting water evaporation and keeping the soil cool. After harvesting the lettuce, you can apply a mulch to the bare soil around the tomatoes.

HERBS

Tomatoes grow well with annual herbs that take up little space, like parsley, dill, and especially basil, and that are planted between them. However, avoid placing perennial herbs like sage, oregano, and rosemary too close. These are more beneficial when planted along the perimeter of the tomato plot as they keep away whiteflies, small-winged insects that transmit diseases to plants, and other pests.

FLOWERS

Tomatoes are frequently combined with French marigolds (*Tagetes patula*), which repel whiteflies and hinder the spread of nematodes in the soil. French marigolds also attract pollinators, resulting in more fruit formation and, therefore, higher yields.

Other annual flowers that attract insect pollinators include calendula, cosmos, nigella, and nasturtium, which are useful and aesthetic companion plants that can cohabit with tomatoes without significantly depriving them of soil nutrients. They can be sown between the plants to avoid blocking the aisles.

You planted your tomato seedlings many days ago, and they are growing well. How much and how often you water, by hand or with an irrigation system, will depend on the weather and plant development. They are already bearing their first flower clusters.

To encourage fruiting, several types of pruning are recommended: sucker removal; pruning to keep two leaders, which improves yields; and pruning lower leaves to promote fruit ripening, while also cleaning up the base of the plants and improving airflow.

Apply fertilizer often but judiciously, to promote fruit development and ripening. The color of your tomatoes is starting to change, a sign that harvest time is just around the corner!

Maintenance, Harvest, and Storage

Irrigation

Tomato plants need about 1 to 2 quarts (1 to 2 L) of water per day, depending on their development stage, your growing methods, and soil quality. The best way to save water and retain soil moisture is to work the soil well, then cover it with mulch (see p. 76).

Before planting tomatoes, make sure to pour 2 to 3 quarts (2 to 3 L) of water into each hole, to build a reserve. Then, plant your seedlings and water them generously. Depending on the soil quality and weather conditions, you should typically wait 6 to 10 days before watering them again to encourage the roots to grow deeper into this pocket of moisture.

After that point, water the plants regularly, not too much and not too little, to avoid blossom-end rot or tomato cracking.

As a rule of thumb, avoid watering tomatoes in full sun and try to prioritize evening and morning irrigation, when the plant's cell tissue is in an active phase, so up to 2 hours or so before sunset.

Irrigation with a Watering Can and Hose

Watering cans and garden hoses are a better fit for home gardeners because they are too time-consuming for professional market gardeners.

Purpose and Advantages

After planting the seedlings, or just before you first water them, build a berm around each plant, to create a watering basin. The berm, 4 to 8 inches (10–20 cm) tall, keeps the water near the plant, preventing it from running off, and allows the water to percolate, making it directly accessible to the roots.

Next, water the plants with a watering can or a hose.

Don't use the sprinkler head on the watering can as it splashes water onto the ground and foliage. As a result, water is lost to evaporation, and leaves are exposed to fungal diseases. Instead, pour straight from the spout into the watering basin.

If you have a big garden, use a garden hose fitted with an adjustable nozzle or wand instead of a watering can, which can be quite tedious. Keep the water pressure low so that the spray won't be too powerful, thus limiting how much water splashes onto the foliage.

Doing It Right

1 Using a hoe, a cultivator hoe, or even just your hand, dig a watering basin around the plant, with a diameter of around 12 to 16 inches (30–40 cm).

2 Slowly pour water into the basin to avoid damaging the berm you just built. Allow the water to gradually percolate through the soil, then repeat the operation to thoroughly wet the ground.

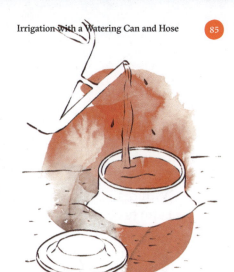

Tip from Jean-Martin Fortier

Even without a drip system, you can use ancient methods to set up reliable and efficient irrigation.

One of the best-known strategies is to take a plastic water bottle, with the cap removed and the bottom cut off, and bury it upside down beside your tomato plant, fill it with water.

Another method involves the use of an *olla*, a type of vessel used in the Roman Empire and up to a thousand years ago in China. The idea is to bury porous clay pots near tomato plants, then fill them with water. Not only do *ollas* slowly release water right by the roots, but they also save 50 to 70 percent on consumption when compared to conventional irrigation. Today, you can find a number of different models adapted to all types of gardens. One drawback is that they are time-consuming to set up: you need to first dig a hole the size of the *olla*, put a layer of sand in it, then place the container in the hole, adding a lid that you tape shut so soil can't get in. Next, fill the hole with fine soil. If it has high clay content, add some sand to improve water transfer and ensure even distribution around the *olla*. If your soil is sandy, add fibrous compost to help increase water retention. Once you have filled the hole, lightly tamp the soil and remove the tape, lift the lid, and fill your *olla* with water.

Note that in clay soil, containers should be buried 2.5 to 3.5 feet (80–100 cm) apart, while in sandy soil, they should be 1.5 to 2 feet (50–60 cm) apart.

Drip Irrigation

With drip irrigation, you can deliver water right at the base of each plant. The system is connected to a timer that controls the watering schedule. This approach is widely used by market gardeners and is gradually gaining support among home gardeners looking to escape the burden of hand watering.

Purpose and Advantages

Drip irrigation distributes water onto the soil, which saves time and water. These systems are typically controlled by an automated timer managing the frequency and quantity of water going to the plants. Unlike micro-sprinklers and sprinklers, drip lines avoid spraying water into the air. This is an ideal setup for tomato crops as it reduces the incidence of disease. In terms of frequency, we recommend watering tomato plants every day, especially once the fruits have begun to set. You can then schedule watering according to the weather: once per day if it's cloudy, twice per day for partial clouds, and three for a particularly hot and sunny day. Ideally, your first watering should happen about 3 hours after sunrise, and the last one should be 4 hours before sunset. To figure out whether you need to water, use a moisture meter, or simply judge the soil's moisture content by poking your fingers into the ground: it should be cool and damp. If the soil is completely dry, it's time to readjust the water supply by increasing its duration or frequency.

Drip Irrigation

Doing It Right

1 In tunnels and greenhouses, install four drip lines, two on each side of the tomato row. Lay a tarp over the lines to reduce water evaporation.

2 In open fields, install drip lines under a mulch or set it on bare soil. You'll be able to see that the water is flowing properly and ensure that each plant is being well watered. However, it will evaporate more quickly on hot days.

Tip from Jean-Martin Fortier

At the start of the season, it's best to use only two of the four drip lines (the ones in the middle of the bed) to avoid leaching nutrients too quickly. These two active lines will keep the soil moist for 2 weeks, then you can open the other two once roots start to develop deeper in the soil and the plants need more water.

Pruning

Pruning is a time-consuming part of tomato care, but it is necessary if you want to maximize yields, harvest good-quality fruits, and prevent the spread of disease. Only prune indeterminate varieties that produce large fruits. Determinate varieties, those that produce smaller tomatoes, and micro dwarf plants require less pruning, sometimes none at all. If you make a mistake, don't panic, tomato plants grow back easily and quickly!

Pruning essentially happens in two steps. The first is removing suckers, which helps the tomato plant focus on developing the main stem and flower clusters, and eventually on ripening fruit.

The second step, once fruits start to form, is removing leaves from the lower part of the plant to prevent cryptogamic diseases (fungal and algal) and—especially—to expose the fruits to sunlight, which helps them ripen.

Pruning Suckers

Pruning suckers means removing young shoots that have developed a stem and a few rows of leaves.

Purpose and Advantages

After they are planted, tomatoes enter an active growth phase. Stems lengthen, leaves get bigger, and the suckers grow too.

These side shoots develop at the leaf axils, diverting and capturing some of the plant's sap, which should be used for fruit development. But even more critical is that letting suckers grow, in the middle of the plant, makes the foliage denser, thus impeding airflow, which in turn fosters the development of cryptogamic (fungal, algal) diseases.

This is why pruning is highly recommended not only to grow bigger fruits, which will ripen faster, but also to clear out the center of the plant, which will limit the spread of disease and provide better access for harvesting tomatoes.

Pruning should be done in dry weather, when the suckers are still young and small. Holding the sucker between your thumb and index finger, break it off in a swift motion; it should detach quite easily. If the suckers have had time to grow and have a thicker stem, you can cut them off with pruning shears or a knife, being careful to make the wound as small as possible.

We recommend pruning every week, or at least every ten days.

Maintenance, Harvest, and Storage

Doing It Right

1 Using your hand or a well-sharpened tool (straight or curved pruning shears, knife, etc.) with a disinfected blade, remove any suckers that have developed at the base of the plants. They are in a prime location to divert resources coming from the roots and must be removed first!

2 Regularly repeat this operation to remove any suckers growing in leaf axils, along the stems. Because they are young and tender, you can easily remove suckers by holding them between your thumb and index finger, then snapping them off in a swift motion.

Tip from Jean-Martin Fortier

When growing tomatoes in the field, you can do without pruning if you don't have enough time, as these plants are more self-sufficient than those grown under shelter. At the start of the season, monitor the crop; if the tomato plants seem to be healthy and growing well, you can ease up on pruning. However, you need to plan for topping them near the end of the season. This involves cutting off the growing tip of each main stem to impede plant growth and direct sap towards growing and ripening the last fruits.

Pruning for Two Leaders

Sometimes it's a good idea not to remove the sucker closest to the base of the tomato plant and let it develop into a second stem, also referred to as the second leader.

Purpose and Advantages

Growing tomatoes with two leaders maximizes your yields from a single plant. However, the fruit on the second leader will make it significantly heavier, and its junction point on the main stem will become quite fragile. To support the plant properly, you need to install a trellis ahead of time, applying the principles used to train tomatoes in greenhouses and tunnels (see p. 74): attach nylon lines to the metal framework of the greenhouse or tunnel every 10 inches (25 cm), then wrap one around each stem. While pruning and training the plants, be careful to avoid shaking them too much. For a two-leader crop, plant seedlings 2 feet (60 cm) apart in the row, a more typical spacing.

Tip from Jean-Martin Fortier

In addition to pruning suckers, you can also remove flowers or flower clusters. If the soil is healthy, keep an average of five clusters; in poorer soil, keep three clusters, or maybe two if you are growing tomatoes in an unfavorable climate (short, humid summers). In tunnels and greenhouses, however, you can keep up to seven clusters.

For varieties that bear larger fruits, we recommend removing the first flower in each cluster, which tends to develop prematurely compared to the others. This will allow the remaining fruit on this branch to develop into uniform tomatoes.

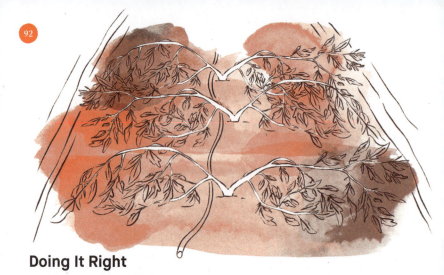

Doing It Right

1 When suckers begin to develop at the base of the plant, select the most vigorous one in the best location to create a strong and healthy junction, then remove the others.

2 To compensate for the weight of the second leader and to avoid putting too much pressure on the junction, set up a sturdy trellis, wrapping nylon line around the stem for support (see p. 74).

Pruning Lower Leaves

For tomato plants, the last type of pruning involves removing the lower leaves, in mid-season, which are often damaged and not productive.

Purpose and Advantages

Once the tomatoes on the lowest fruiting branches begin to ripen, prune the lower leaves, which often turn yellow and host fungal diseases. This process also improves airflow around the base of the plant, while allowing light and sunshine to reach the fruit and contribute to ripening. Start at the bottom, then work your way up to the first fruit cluster. Repeat this operation every 2 weeks or so.

Doing It Right

 By hand or with a sharp disinfected blade, cut the leaf branch as close as possible to the main stem.

Tip from Jean-Martin Fortier

If left unpruned, tomato plants that bear medium-to-large fruit will grow vigorously, which is ultimately detrimental to fruit production and overall plant health. However, those that produce smaller fruit, like currant, cherry, and cocktail tomatoes, will deliver better yields if left alone.

Fertilization

Tomatoes are heavy feeders, requiring nutrients for their stems and leaves, as well as their fruits. The soil will naturally provide some, generated by plant matter that has decomposed to become humus, but it's not enough to meet the needs of tomatoes throughout their lifespan. It is therefore essential that you add fertilizer, organic when possible. These fertilizers, either plant based (compost, liquid manure, castor meal, etc.) or animal based (manure, chicken manure, etc.), supply nutrients and organic matter to the soil and boost the many microorganisms that ingest plant matter to transform it into humus.

Tomato plants can be fertilized in two ways. The first, initial fertilizing, occurs during soil preparation (see p. 55). The second described here are inputs added approximately 6 to 8 weeks after planting the seedlings. This fertilizer replenishes the initial nutrients consumed by the plants and provides the nutrients needed to develop fruits.

Mid-Season Fertilization or Side Dressing

As they grow, tomato plants need additional fertilizer rich in potassium, which is essential for fruit development resulting in high-quality, flavorful tomatoes.

Purpose and Advantages

To get the most out of fertilizers and maximize their effect on tomatoes, you must monitor your plants, noting the appearance of the fruits and, especially, the leaves. These are valuable indicators of the plant's nutritional needs. Deficiencies in nitrogen, phosphorus, potassium, and trace elements can be detected simply by analyzing the foliage. The next step is to add the right kind of fertilizer. Fertilizer packaging typically features the letters "NPK" (for nitrogen, phosphorus, and potassium), followed by a number indicating the percentage of each component. For example, 15-5-10 NPK means that it contains 15% nitrogen, 5% phosphorus, and 10% potassium.

→ Nitrogen Deficiency

Foliage turns pale green and yellowish, or drops, and the plant stops growing. In response, apply castor meal (NPK 5-2-1, dosage: 1.6 oz./ft.^{2}, 0.5 kg/m^2) around the base of the plant, or dilute the following into your irrigation water: seaweed solution (NPK 2-1-0.5, dosage: 0.2 to 0.4 cup/ft.^{2}, 0.5 to 1 L/m^2); blood meal (NPK 10-1-0, dosage: 1 oz./ft.^{2}, 0.3 kg/m^2); and beet molasses (NPK 4-2-7, dosage: 0.2 to 0.4 cup/ft.^{2}, 0.5 to 1 L/m^2). To address this imbalance in the long run, amend the soil with compost or manure.

→ Phosphorus Deficiency

Foliage has a blue-green tint, wilts or drops, and plant growth slows down. In response, apply bone meal (NPK 2-20-0, dosage: 0.7 oz./ft.², 0.2 kg/m²) and wood ash around the plants, then water the soil. To address this imbalance in the long run, amend the soil with compost and improve soil drainage.

→ Potassium Deficiency

Foliage and stems turn brown and become necrotic, and plant growth slows. In response, dilute the following in water and apply every 2 to 3 weeks: nettle or comfrey tea and beet molasses (NPK 4-2-7, dosage: 0.2 to 0.4 cup/ft.², 0.5 to 1 L/m²). To address this nutrient imbalance in the long run, amend the soil with manure, potassium chloride (also known as potassium salt), and wood ash.

→ Magnesium Deficiency

Foliage turns yellow (veins remain green), then falls off. This is known as interveinal chlorosis. In response, spray the leaves with a 2% magnesium sulfate solution. To address this nutrient imbalance in the long run, add dolomitic lime or kieserite powder.

→ Iron Deficiency

Foliage and new growth turn yellow, then become bleached. Plant growth slows and can even stop. This is known as iron chlorosis. In response, spray the leaves with a 0.3% iron sulfate solution, which is sometimes called anti-chlorosis treatment. To address this nutrient imbalance in the longer run, increase soil acidity and amend it with compost.

Tip from Jean-Martin Fortier

To grow healthy tomato plants, it is essential that you establish a fertilization schedule and stick to it, using the quantities recommended on commercial packaging. Applying too much fertilizer will weaken the tomato plants and make them vulnerable to disease and insect damage. For example, excess nitrogen softens tomato stems and leaf tissue, making them more vulnerable to aphids. Too much manure results in wilting foliage and the plant's progressive decline.

To remedy this, flush the soil with water to wash out nutrients, and prune or cut back the plant to give it a boost.

Mid-Season Fertilization or Side Dressing

Doing It Right

1. Reduce compaction and loosen the soil with a hoe, so it will be easier to incorporate the fertilizer.

2. Over moist soil, spread organic fertilizer (decomposed manure, castor meal, etc.) around the base of the plant, then incorporate it into the soil and water the crop to help the fertilizer dissolve and release nutrients.

3. If using a tomato-specific pelleted fertilizer, spread one or two handfuls around the base of each plant, using the dosage recommended by the manufacturer, then water the plant.

If using liquid fertilizer, dilute it in your watering can and pour over the base of the plant, after creating a watering basin, to keep it closer for the roots to absorb.

Types of Fertilizers

Fertilizers can be organic, i.e., derived from organic matter (plant or animal), or chemical (industrially manufactured). They contain three main elements in different proportions: nitrogen (N), which is needed for stem and foliage growth; phosphorus (P), which helps with root, flower, and fruit development; and potassium (K), which is essential for storing nutrients and adding color to vegetables. Many trace elements (iron, zinc, copper) are present in smaller amounts and play an active role in helping the plant assimilate the three main elements (NPK).

→ ANIMAL-BASED ORGANIC FERTILIZERS
To grow tomatoes, composted horse manure or cattle manure is especially recommended. In light soils, it helps retain water while improving soil quality.

Other animal-based fertilizers can be used as alternatives to this manure. These include chicken manure, guano, blood meal, bone meal, horn meal, feather meal, and fish meal.

They act faster, and roots access their nutrients more quickly. They can be applied directly onto the soil around the plants, or, in some cases, they can be dissolved into water used for irrigation.

→ PLANT-BASED ORGANIC FERTILIZERS
These fertilizers provide nutrients for plants and strengthen their natural defenses.

They are packaged as powders (meals) and as liquid solutions (molasses or teas). Some, like nettle and wormwood, are insect repellents, while others, like rhubarb or horsetail, strengthen plant tissue to help it withstand cryptogamic diseases (fungal, algal). As for wood ash, it provides potassium.

To apply these fertilizers, spread powders over the soil and dilute molasses and teas into water, then apply either through irrigation or as a foliar spray.

Fertilizer Teas

To give tomatoes (and other vegetables) a boost, you can use certain plants, wild or cultivated, that act as insect repellents, fungicides, and even fertilizers. Beneficial molecules can be extracted from these plants through various processes: fermentation, maceration, decoction, and infusion. Foliage, stems, and flowers are the basic ingredients in these different solutions. To more easily extract the beneficial molecules, chop or crush the plants before following the steps described below, depending on the solution you are preparing.

→ **MAKING YOUR OWN FERMENTED PLANT-BASED TEA**
1 • In a container of rainwater, soak your chopped plant matter, maintaining a ratio of 13 ounces of plant matter to 1 gallon of water (1 kg/10 L). Stir regularly during the fermentation process, which lasts 5 to 30 days. The liquid is ready once it looks like tea and no more bubbles appear when you stir it.
2 • Filter the liquid and store it in bottles. The tea will keep for several months in a cool, dark room. For a foliar spray, dilute the tea to a ratio of 1:4 (tea: water).

→ **MAKING PLANT-BASED TEA FROM A MACERATION**
1 • Macerate 2.2 pounds (1 kg) of chopped leaves in 2.5 gallons (10 L) of rainwater, stirring daily. Fermentation lasts from 2 to 10 days.
2 • Filter and use for watering, undiluted, or as a foliar spray with a 1:4 dilution (tea: water).

→ **MAKING PLANT-BASED TEA WITH BOILING WATER**
Option 1 — Decoction (boiling down):
1 • Macerate 3.5 ounces (100 g) of fresh leaves in 1 quart (1 L) of water for 1 day, then boil it for 20 to 30 minutes.
2 • Let the mixture cool, then filter the liquid, and use it as an undiluted spray.

Option 2 — Infusion:
1 • Drop 3.5 ounces (100 g) of chopped plant matter into 1 quart (1 L) of boiling water.
2 • Let the mixture cool, filter the liquid, and immediately use it undiluted.

Harvest

For tomato growers, the last step is the harvest, when the fruits have reached maturity. Ripe tomatoes can be broken down into two categories: ready-to-pick and ready-to-eat.

In the first stage, ready-to-pick, the color looks right, but the flesh is still a little firm. This allows growers, especially professional market gardeners, to handle the tomatoes without damaging them before they can be sold.

In the next stage, ready-to-eat, tomatoes have a deeper coloring, and more importantly, the flesh feels a little soft when you give it a gentle squeeze.

Tomato plants tend to reach these stages of maturity in late June or early July, depending on whether they are grown under cover or in the field, and harvests will continue until the first fall frosts.

Once the fruit is picked, its quality quickly declines. That's why tomatoes should be eaten as soon as possible or preserved. Tomatoes can be processed in many ways that include canning, drying, and making sauces and juices.

Harvesting Tomatoes

After a long process in which the tomatoes change color and their flesh ripens—a process that makes them tasty—it's harvest time. This is the final stage of tomato growing.

Purpose and Advantages

Harvest dates will vary, depending on the variety. Tomatoes ripen and reach their final color in the last few growing days.

This transformation in the fruit is caused by a buildup of sugars, which the plant produces through photosynthesis.

To promote more sugar production in tomatoes, make sure the plants have good sun exposure and are bearing a reasonable number of fruits. Since color is a good indicator of ripeness, it's best to wait until it has reached the color described by the seed producer.

Once ripe, tomatoes begin to soften: this is a sign that they are ready to be harvested.

Tip fromJean-Martin Fortier

To promote fruit development towards the end of the season, about 6 weeks before the last scheduled harvest, prune all emerging flower clusters. It's also time to top all plants, pruning all stems above the last cluster of pollinated flowers.

Maintenance, Harvest, and Storage

Doing It Right

1 When hand picking, start by visually identifying fruits that are ripe enough.

With classic tomatoes, gently grasp the fruit from below, then press your thumb or index finger onto the knuckle on the stem just above the fruit, which is a natural breaking point.

With cherry tomatoes, simply pull on each fruit to separate it from the truss.

Once harvested, gently place them in a crate or basket.

2 Harvesting with pruning shears makes it easier to separate the fruit from the stem with a clean and fast cut. Gently grasp the ripe tomato and cut the stem just above the fruit to separate it from the truss. Place the harvested tomatoes side by side in a basket or crate, in a single layer.

Temporarily Storing Tomatoes

Since tomatoes are delicate, handling and storage are short-term, transitional steps that occur before they are sold and eaten.

Purpose and Advantages

Once you've harvested tomatoes, sort them to remove any misshapen or diseased fruits, especially those affected by late blight and early blight. Then, clean them and store in reusable crates in a cool area.

Tomatoes don't need to be kept in a cold room, just in a cool one, around 54°F to 55°F (12°–13°C). Because tomatoes quickly lose their flavor, you must store them for as little time as possible, a few days at most.

Doing It Right

1 Professional market gardeners need to clean and dust off ripe tomatoes before storing them in reusable crates. At this stage, they can also be gently washed in a bubbler. After drying them, weigh the fruits and store in a cool place until they go to market.

Storing Tomatoes at Home

For home gardeners, it's often difficult to eat all the fresh tomatoes after harvesting. They can be preserved in various ways to avoid any going to waste.

Purpose and Advantages

There are many ways to preserve your tomato harvest. They can be made into sauces, dried, or cooked and stored in jars. The advantage is that you'll be able to eat tomatoes all winter long!

At the end of the season, you can also harvest the last green tomatoes and let them ripen gradually. However, they will have a slightly milder flavor.

Tip from Jean-Martin Fortier

At the end of the season, temperatures are sometimes too low for the last well-developed fruits to ripen and turn red. At this point, you can pull out the tomato plants and hang them in a warm spot, such as a veranda or tunnel. The most important thing is to keep the space warm; it doesn't need to be especially bright. The tomatoes will slowly change color, and you can eat them.

Doing It Right

1 → SHORT-TERM STORAGE
When tomatoes are picked green or not quite ripe, they can be kept in a dry place for up to a few weeks. It is best to store them in a single layer, leaving a little space around each tomato. Monitor them closely, to prevent the spread of any mold.

2 → LONG-TERM STORAGE
Several processing methods can be used to preserve tomatoes for more extended periods of time.

Small to medium-sized varieties can be dried, salted and preserved in jars, or pickled.

Cooking is also an effective method, allowing you to enjoy your harvest in coulis, juices, and concentrates stored in sterilized jars or even in containers suitable for freezing.

Since tomatoes are quite susceptible to disease, pests, and physiological disorders, the list of known ailments affecting them is very long and ever evolving. Only the most common issues are mentioned in this book.

By applying good growing practices, you will significantly limit their development. Many actions contribute to preventing pest damage in tomatoes, including using tunnels and greenhouses, proper watering, regular plant care, and crop rotations. In terms of disease, when the first symptoms appear, you need to assess the extent of the damage and decide whether it requires treatment. As a last resort, if the situation is desperate, you may have to destroy the infested plants to save the rest of the crop.

When it comes to parasites, many preventive measures can limit their spread, such as the use of companion plants (see p. 79) for outdoor tomatoes, as well as pheromone traps and predatory insects for those grown in tunnels and greenhouses.

Tomato Enemies

This section covers some of the more common diseases and pests that affect tomatoes in temperate regions. Depending on your location and climate, there may be other enemies you need to research and manage.

Diseases

Tomato diseases are caused by fungi or viruses, which are often transmitted by parasitic insects.

Downy Mildew (late blight)
→ Symptoms
Of all the fungal diseases affecting tomatoes, downy mildew, also called late blight, is the most widely known. It is caused by *Phytophthora infestans*, which overwinters in the leaves and stems and can also be present in the soil. It is highly contagious and is especially prevalent at the end of the season and in wet summers. It spreads when temperatures are cool, between 59°F and 77°F (15°–25°C), and humidity is high. Plants hit by downy mildew develop irregular brown spots on foliage and young fruit, as well as gray fuzzy growth on the underside of leaves. When only partially affected, tomatoes are still edible after the damaged parts are removed.

→ Preventive Measures and Disease Control
Downy mildew is less likely to affect tomatoes grown in tunnels or greenhouses if there is good airflow around the plants and the soil is mulched. Similarly, you can reduce the risk of it developing by watering tomato plants at the base and not on the leaves, implementing a crop rotation, and clearing all plant debris from the soil. For cases of recurring downy mildew, you may opt for tomato varieties that are disease resistant. As a preventive measure, you can apply a nettle or comfrey tea, or even a garlic or horsetail decoction (concentrate). Spray affected plants with a copper sulfate solution.

Diseases 109

Early Blight
→ Symptoms
While early blight is often mistaken for downy mildew (late blight), it is much more harmful. It is caused by fungal species belonging to the Alternaria genus, including *A. solani*, *A. alternata*, and *A. tomatophila*, that can damage tomato plants. It typically occurs from July to November, during hot and humid stretches, especially in wet weather. It presents as well-defined dark brown spots with concentric rings, first in the leaves closest to the ground before eventually spreading to the entire plant. Once affected, tomatoes rot quickly. As soon as symptoms appear, cut off and burn the infected parts of the plant.

→ Preventive Measures and Disease Control
To prevent early blight, mulching is a good practice, as are crop rotation and irrigation methods that water the base of the plant without wetting the foliage.

You can also spray plants with a lime milk solution to destroy any spores on the main stem. Another good prevention measure is to apply a decoction of garlic or horsetail through a foliar spray. Lastly, once spots appear, you can spray the plants with a copper sulfate solution.

Botrytis (gray mold)
→ **Symptoms**

This common disease is caused by the fungus *Botrytis cinerea*. It appears when humidity is high, temperatures are moderate, airflow around the plants is limited (especially in tunnels and greenhouses), and too much nitrogen was added in fertilization. It affects the entire plant, from the collar to the foliage, including stems and fruits. Affected foliage turns yellow, dries out, and drops; a gray fuzz develops on the flowers and fruits.

→ **Preventive Measures and Disease Control**

To prevent botrytis, you can limit nitrogen inputs, plant garlic within tomato rows, spray the base of each plant with lime milk, and apply a horsetail decoction (concentrate). To combat this fungus, the best treatment is a foliar spray with a sulfur-based solution.

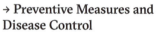

Powdery Mildew
→ **Symptoms**

Powdery mildew is a fungal disease that appears as a white and powdery growth on the plant. It spreads quickly in the summer, when hot days follow cool damp nights. Symptoms include wilting foliage, dropping young fruit, cracking in older fruits, and overall stunted plant growth. Affected parts of the plant should be removed and destroyed.

→ **Preventive Measures and Disease Control**

The simplest way to prevent the spread of powdery mildew is to increase the distance between plants and improve airflow, apply lime milk to the base of the plant, limit the amount of nitrogen in fertilizer inputs, and apply a horsetail or nettle tea. If contamination persists from one year to the next, a good solution is to opt for disease-resistant varieties. To control the disease, spray the plant with a sulfur-based solution.

Verticillium Wilt
→ Symptoms
Verticillium wilt is a disease caused by the fungus *Verticillium dahliae*, which prevents sap from circulating properly in tomato stems. It causes leaves to wilt, turn yellow, and fall off.

→ Preventive Measures and Disease Control
There is no cure for this disease. Therefore, you need to cut off and destroy damaged parts of the plant as soon as symptoms appear.

ToBRFV: An Emerging Virus!

A new disease has appeared in recent years: the tomato brown rugose fruit virus (ToBRFV). Affected tomatoes show yellow or brown spots, become rough, and can no longer be eaten. The virus causing this damage is transmitted through simple contact. There is currently no treatment. At this time, however, only industrialized farms with large-scale monocultural practices have been affected. If ToBRFV is found in a crop, all plants in the affected area must be uprooted and destroyed.

Insect Pests and Parasites

Some pests and parasites damage tomatoes. Larvae, butterflies, moths, and flies cause irreversible damage. So it's best to protect yourself before an infestation occurs.

Cutworms (moth caterpillars)
→ Symptoms
In their larval form, moths, also known as cutworms, grow up to 2 inches (5 cm) long and devour fruit and leaves overnight. They particularly like leaf blades, the thin flat part of the leaf, and stems near the soil line. During the day, they take shelter in tunnels they dig in the ground.

→ Preventive Measures and Disease Control
As a preventive measure, you can attract birds, which are fond of moths, by planting berry bushes or installing nesting boxes, for example, near the tomatoes. You can also use mulch to keep your soil cool and hinder the progress of cutworms, as they dislike damp soil. Lastly, you can set up yellow sticky traps to catch the moths, and place tiles or wood planks near the plants, which cutworms use for shelter overnight. Early in the morning, you can collect them.

To deal with an invasion, you can apply a decoction of tansy or wormwood or a solution made from a walnut leaf maceration to the base of infested plants. You can also spray a solution containing *Bacillus thuringiensis* (Bt or BtK), which kills caterpillars by attacking their digestive systems. In greenhouses and tunnels, you can even introduce another cutworm predator, the *Steinernema feltiae* nematode, to get rid of it.

Whiteflies
→ Symptoms

Whiteflies, or *Trialeurodes vaporariorum*, are small white flies that enjoy tomato foliage. Adult flies lay their eggs on the undersides of leaves. Once they hatch, the nymphs feed on plant sap and secrete honeydew, which is then colonized by a fungus called sooty mold. Next, a blackish film begins to cover the leaves, and plant growth slows. Whiteflies mainly appear in greenhouses and at the end of the season. Around that time, we recommend inspecting the foliage regularly to detect the first colonies and act quickly.

→ Preventive Measures and Disease Control

To prevent an infestation, remove any weeds, as some may be host plants for whiteflies, and install yellow sticky traps to stop adult flies and monitor populations so you can decide any need for treatment. To deal with an infestation in a tunnel or greenhouse, you can introduce natural predators like the parasitic wasp *Encarsia formosa*, ladybugs (*Delphastus pusillus*), and the fungus *Lecanicillium muscarium* (syn. *Verticillium lecanii*). You can also apply a decoction of tansy or tobacco, or a nettle or wormwood tea. One last solution is to use a soft soap foliar spray.

Slugs
→ Symptoms
Although slugs prefer greens, they can also attack tomatoes by severing the stems of newly planted seedlings or taking refuge in tunnels dug into fruits (to hide from predators). In vegetable gardens, several slug species can be found, including *Limax aggtis*, *L. maximus*, and *Arion hortensis*. They are active at night, when it is humid and temperatures are above 50°F (10°C).

→ Preventive Measures and Disease Control
The easiest way to prevent slug damage is to encourage their predators like hedgehogs, amphibians, and birds. To capture slugs, you can lay a tile on the ground, under which they will shelter, or set a trap filled with beer, which slugs love. A nonalcoholic beer is preferable, to avoid harming other animals, such as hedgehogs. Lastly, you can spray the plants with a solution made from a blackcurrant or fern maceration.

Nematodes
→ Symptoms
Nematodes are microscopic worms that live in plant tissue. They carry diseases as they travel from one plant to another. Affected plants become misshapen, turn yellow, and eventually wilt. The most dangerous nematode for tomatoes, the root-knot nematode, feeds on the roots, where it generates bumps called galls, and can lead to the total dieback of the plant. Nematodes are difficult to manage and spread easily. They are most often found in greenhouses. Once the soil is contaminated, it can be a challenge to eradicate these parasites, unless you stop growing tomatoes in that area for the next 4 to 5 years.

→ Preventive Measures and Disease Control
Although there is no treatment for nematodes, many preventive measures can help, like crop rotation and the destruction of all contaminated plants. You can also grow tomato companion plants (see p. 79) like marigolds, which emanate a strong odor that is a powerful nematode repellent.

Insect Pests and Parasites

Tomato Leafminer (*Tuta absoluta*)
→ Symptoms

The tomato leafminer, also known as *Tuta absoluta* and *Phthorimaea absoluta*, is a moth belonging to the Lepidoptera order that can wreak havoc on tomato plants. Native to South America, it was first found in Europe in 2006 and has since spread to the Middle East, parts of Africa, South Asia, and Central America. It has not yet become established in North America. The larvae mine into the leaves, stems, and ripe or green fruits, leaving behind whitish blotches. Leaves become necrotic, and fruits rot, rendering them completely unfit for consumption.

→ Preventive Measures and Disease Control

All damaged parts of the plant must be destroyed, including debris remaining in the soil, such as roots or stem fragments. To manage an infestation, spray the plant with an insecticide containing *Bacillus thuringiensis* (Bt). Preventively, and only in covered spaces, install insect netting along any openings to keep leafminers out. However, this will also stop pollinators from entering your greenhouses and tunnels.

Tip from Jean-Martin Fortier

Crop monitoring is your greatest weapon in the fight against pests and parasites. The best approach is to thoroughly inspect tomato seedlings every week, even using a magnifying glass, if necessary. Once a pest has become established, you need to react quickly and hit it hard because they multiply fast and will destroy your crop in no time. In the event of an infestation, you can use multiple approaches to address the problem: introducing predators, setting up pheromone traps, installing insect netting. However, these processes are better suited to enclosed spaces (greenhouses or tunnels) than open fields.

Physiological Disorders

Physiological disorders are problems caused by environmental stresses and unsuitable growing conditions. They prevent tomato plants from growing and functioning properly.

Blossom-End Rot
→ Symptoms
In addition to splitting, blossom-end rot is the best-known physiological disorder affecting tomatoes. Because the bottom of the fruit turns black, it is sometimes called bottom rot. This affects green fruit that has already begun to grow, and can be an indicator of calcium and magnesium deficiencies. It also occurs when soil is too dry, watering is irregular, or plants are overloaded with fruit. The lesions caused by blossom-end rot are a vector for fungal diseases.

→ Preventive Measures and Disease Control
There's no treatment for blossom-end rot, but regular, consistent watering is a simple way to avoid this disorder. You can also prevent it by applying a calcium-rich nitrogen fertilizer (like bonemeal) at the base of the plant, as well as a few handfuls of sand, to improve soil aeration and encourage microorganism activity.

Puffiness
→ Symptoms
Sometimes, tomatoes develop into a triangular or heart shape, with hollow cavities instead of flesh and seeds; this is tomato puffiness. This defect appears when temperatures are too high (over 95°F, 35°C) or too low (below 53° to 57°F, 12° to 14°C), and when pollination is insufficient or of poor quality, watering is inconsistent, and inadequate fertilizer was added.

→ Preventive Measures and Disease Control
To manage this problem, ensure regular watering, occasionally supplemented with comfrey tea, and avoid fertilizing with too much nitrogen, instead opting for potassium-rich fertilizers.

Splitting
→ Symptoms
As the fruit ripens, it sometimes develops cracks that start around the stem end (peduncle). These cracks may expand, developing into radial splitting down the fruit or into concentric circles at the top. Tomatoes may even burst. This cracking is caused by temperature fluctuations and inconsistent watering, which can also affect plant growth. Some varieties are more prone to cracking.

→ Preventive Measures and Disease Control
Mulching the soil and installing an automatic watering system are the best solutions to this problem. You can also apply more fertilizer, improve airflow in your greenhouses and tunnels, and choose varieties that are less prone to cracking.

Discoloration and Incomplete Ripening

If tomato plants are overfertilized or exposed to environmental stressors (low temperatures), a part of the fruit, often the shoulders, may remain green or turn yellow. Affected areas will not ripen. In heavily affected plants, the flesh becomes white and hard, even though the skin appears red. This problem is mostly seen in tomatoes grown outdoors, and some varieties are more susceptible to the disorder.

Partial ripening is another type of discoloration that affects both the skin, which develops green, yellow, and red mottling, and the flesh, which has gray and yellow patches that do not ripen. This physiological disorder can be caused by a boron or potassium deficiency, excess nitrogen, high humidity (especially in tunnels and greenhouses), significant temperature variations, and excessively compacted soil.

Stitching and Catfacing

With tomato stitching, or zippering, scars run from the stem to the bottom of the fruit and look like a stitched cut. Holes may also develop along this scar. These symptoms only appear once the fruit has reached maturity. They are still fit for consumption, provided the browned part is removed.

Some fruit becomes misshapen because of unfavorable growing conditions (extreme temperatures and high humidity) during the flowering and pollination stages. This is true for catfacing, which appears when temperatures fall below 59°F (15°C) during the flowering stage. Brown scars and cavities appear on the blossom end of the fruit. Catfacing can also be caused by excess nitrogen.

Physiological Disorders

Leaf Curl

If the weather is hot and dry, or the soil is waterlogged, the tips of the leaves may curl inward, and their texture may become fibrous. This behavior is called leaf curl. It is worth noting that in some varieties, leaves do this to conserve moisture. In such cases, the leaf curl does not affect tomato yields.

Tomato Blossom Drop

Flowers may fall off the plant when exposed to the following conditions: high temperatures (over 86° to 95°F, 30° to 35°C), excessive humidity, drafts (especially in greenhouses and tunnels), nutrient deficiencies, or the presence of diseases or parasites. The flower or bud, leaf stem (petiole), and young developing fruit turn yellow, then brown, and eventually dry out and fall off the flower cluster. Some varieties are more prone to leaf curl than others.

Tip from Jean-Martin Fortier

In the early days after transplanting tomato seedlings, we recommend taking preventive measures, like applying a nettle or comfrey tea every 10 days.

As soon as the first flower clusters appear, spread a spoonful of lithothamnion on damp soil at the base of the plant to prevent blossom-end rot.

Later, in early August, you can water the seedlings with a horsetail tea solution to help strengthen the foliage's resistance to disease. It's an excellent alternative to using Bordeaux mixture fungicide, which can be detrimental to soil life.

Acknowledgments from Jean-Martin Fortier

I wish to thank the entire team at the Market Gardener Institute for encouraging me to pursue my mission, every day. A big thank-you also goes out to the Growers & Co. team, who pushes me to come up with new types of equipment! I especially want to acknowledge my partner, Maude-Hélène Desroches, who is an exceptional market gardener and a dear friend!

Acknowledgments from New Society Publishers

We extend a great thanks to Delachaux et Niestlé, the French publisher, for working with us to publish this English edition. Further thanks to the New Society Publishers team for producing the book and especially to Laurie Bennett for her meticulous attention to technical details and high-quality translation into English.

Acknowledgments from Delachaux et Niestlé

A big thank-you to Jean-Martin Fortier and his team at the Market Gardener Institute for this wonderful collaboration.

Our heartfelt thanks go out to Pierre Nessmann for putting us in touch with Jean-Martin, for thoroughly editing this collection, and for being so generous with his time. For this book, we owe him so much. We also wish to thank Flore Avram, whose illustrations give this collection a beautiful, simple character; to Grégory Bricout for graphic design that cleverly reflects Jean-Martin Fortier's spirit; and to Sandrine Harbonnier and Sabine Kuentz for their work on the text.

Reference Books

The Market Gardener: A Successful Grower's Handbook for Small-Scale Organic Farming, New Society Publishers, 2014.
Microfarms: Organic Market Gardening on a Human Scale, New Society Publishers, 2024.

Grower's Guides from the Market Gardener

Tomatoes: A Grower's Guide
Vegetable Garden Tools: A Grower's Guide

Coming Soon

Root Vegetables: A Grower's Guide
Living Soil: A Grower's Guide
Fruit Vegetables
A Year of Vegetables

Translator : **Laurie Bennett**

About New Society Publishers

New Society Publishers is an activist, solutions-oriented publisher focused on publishing books to build a more just and sustainable future. Our books offer tips, tools, and insights from leading experts in a wide range of areas.

We're proud to hold to the highest environmental and social standards of any publisher in North America. When you buy New Society books, you are part of the solution!

At New Society Publishers, we care deeply about *what* we publish — but also about *how* we do business.

- This book is printed on **100% post-consumer recycled paper**, processed chlorine-free, with low-VOC vegetable-based inks (since 2002)
- Our corporate structure is an innovative employee shareholder agreement, so we're one-third employee-owned (since 2015)
- We've created a Statement of Ethics (2021). The intent of this Statement is to act as a framework to guide our actions and facilitate feedback for continuous improvement of our work
- We're carbon-neutral (since 2006)
- We're certified as a B Corporation (since 2016)
- We're Signatories to the UN's Sustainable Development Goals (SDG) Publishers Compact (2020–2030, the Decade of Action)

To download our full catalog, sign up for our quarterly newsletter, and to learn more about New Society Publishers, please visit newsociety.com.

ENVIRONMENTAL BENEFITS STATEMENT

New Society Publishers saved the following resources by printing the pages of this book on chlorine free paper made with 100% post-consumer waste.

TREES	WATER	ENERGY	SOLID WASTE	GREENHOUSE GASES
14 FULLY GROWN	1,100 GALLONS	6 MILLION BTUs	49 POUNDS	6,300 POUNDS

Environmental impact estimates were made using the Environmental Paper Network Paper Calculator 4.0. For more information visit www.papercalculator.org

MIX
Paper | Supporting responsible forestry
FSC® C016245

new society
PUBLISHERS
www.newsociety.com